SpringerBriefs in Ethics

More information about this series at http://www.springer.com/series/10184

Garðar Árnason

Foucault and the Human Subject of Science

 Springer

Garðar Árnason
Institute of Ethics and History of Medicine
University of Tübingen
Tübingen, Germany

ISSN 2211-8101   ISSN 2211-811X   (electronic)
SpringerBriefs in Ethics
ISBN 978-3-030-02812-1   ISBN 978-3-030-02813-8   (eBook)
https://doi.org/10.1007/978-3-030-02813-8

Library of Congress Control Number: 2018960284

This Springer imprint is published by the registered company Springer Nature Switzerland AG
The registered company address is: Gewerbestrasse 11, 6330 Cham, Switzerland

For Sandra, Lona, and Elisa

Acknowledgements

This book has been taking form over a long time. It grew out of my doctoral thesis, parts of which survive in this version. I remain grateful to the supervisor of my thesis, the late André Gombay, and the reader, Ian Hacking, for extensive discussions and advice over the years. Various other people have read and commented on earlier drafts of work that ended up in this book. I am particularly grateful to Donald Ainslie, Vilhjálmur Árnason, Mark Kingwell, Robert Mugerauer, Amy Mullin, Skúli Sigurðsson, and Ingrid Stefanovic. They do not necessarily share my views, and any inaccuracies, errors, and shortcomings are my responsibility alone.

Parts of this book have been presented at various conferences and meetings. A paper based on Sects. 2.1 and 2.2 was presented at the workshop "Rationality and Social Predictions," which was organized by the NorFA Network "Local and Global Modes of Scientific Rationality" and held in Tromsø, Norway, in 2004. A paper based on Chap. 3 was read at the Institute for Science, Ethics and Innovation (ISEI), University of Manchester, UK, in 2008. Parts of Chap. 3 were also presented at the 25th European Conference on Philosophy of Medicine and Health Care in Zürich, Switzerland, in 2011. A paper based on Chap. 4, the study of left-handers, was presented at the 28th European Conference on Philosophy of Medicine and Health Care in Debrecen, Hungary, in 2014. Parts of Chap. 4 appear in "Biopolitics and the Longevity of Left-Handers" in Péter Kakuk (ed.) *Bioethics and Biopolitics* (Springer), pp. 59–76. Parts of Chap. 5 on the Icelandic Health Sector Database have been presented at the conference of the Society for Philosophy and Technology (SPT) in Aberdeen, Scotland, in 2001; and at the 18th European Conference on Philosophy of Medicine and Health Care in Reykjavík, Iceland, in 2004. I thank audiences and participants at these events for very useful discussions. Parts of Sects. 5.3 and 5.4 have appeared in a book chapter: "Interbreeding within the Icelandic population is high compared to that of mice or fruit-flies", in G. Árnason, S. Nordal and V. Árnason (eds.): *Blood and Data: Ethical, Legal and Social Aspects of Human Genetic Databases* (Reykjavík: University of Iceland Press, 2004). I thank the University of Iceland Press for the permission to reprint material from that chapter.

For their support and inspiration over the years, I thank Vilhjálmur Árnason, Matti Häyry, Doris Schroeder, Tuija Takala, and Urban Wiesing. Finally, I thank three anonymous reviewers for very helpful and constructive comments and suggestions.

Contents

Chapter 1
Introduction

Scientific knowledge has all sorts of effects on us. It can, for instance, change how we see ourselves, our relations to each other and how we relate to the environment. We can call these effects of scientific knowledge "power effects," and we can call scientific discourses "discourses of truth"—in so far as scientific knowledge makes a greater claim to truth than anything else in our times. When we are under pressure to change how we see ourselves, our relations to each other or how we relate to the environment, then it may very well be the case that this pressure gets its force from the authority of science as a discourse of truth. Sometimes we are subjugated by a discourse of truth, perhaps because we happily, apathetically or uncritically accept its authority and power, perhaps because we just do not know how to resist it. We may get particularly worried if the scientific knowledge in question is knowledge about us, or about a set of individuals to which we belong.

Michel Foucault once proposed a critique of power effects and discourses of truth. He described this critique as "the movement by which the subject gives himself the right to question truth on its effects of power and question power on its discourses of truth" (Foucault 1997 [1978], p. 32). I shall take up his idea of a critique, develop it a little further and then apply it to two cases. I call this sort of critique "politics of truth," a label that has to be understood in its Foucauldian context. This book is therefore about politics of truth and at the same time an example of a politics of truth. My aim is to offer a way of resisting science, by exactly questioning discourses of truth and their power effects.

But why would one need to resist science? One might object that the sciences simply discover things about the world, things that we as rational beings either believe or else stubbornly close our eyes to. Criticizing a science, or a piece of scientific knowledge, or a certain scientific practice, would then be equivalent to shooting the messenger, at least in so far as the science is sound. The only justifiable criticism would be that the science in question is dubious, that it had done a bad job of finding out how things really are. Perhaps because it has failed to exclude other values than strictly cognitive ones. If the scientist is working properly, he will only let cognitive values influence his results and not his political, social, moral or religious

© The Author(s) 2018
G. Árnason, *Foucault and the Human Subject of Science*, SpringerBriefs in Ethics,
https://doi.org/10.1007/978-3-030-02813-8_1

values. In other words, when science works properly, the scientist simply reports what he has discovered. The politics of truth, which I here discuss and participate in, favors another picture of science. Politics of truth pictures science as a struggle, for instance:

(1) when scientific knowledge is produced (involving scientists struggling with each other as well as with administrators, grant agencies, science policy makers, interest groups, the media and, in some cases, the scientific subjects);
(2) when scientific knowledge is demanded or used in order to exert power; and
(3) when scientific knowledge has effects of power, without anyone consciously using it for some end or another.

Science is involved in power struggles at many levels and in different parts of society, but these struggles are not only external to science: power struggles also take place within science. These power struggles do not necessarily corrupt scientific practice or lead to falsity, although they can. They may even be necessary for the success of science by providing motivation, pressure and opportunities. Not only do they shape the external relations of science, the very form and content of scientific knowledge is constantly being negotiated through these struggles. Science is sometimes political, in the sense that it sometimes gets caught in partisan politics or is debated in parliaments—for instance in debates about science policy or in debates about the nature and weight of scientific evidence used in policy-making, or even in debates about what science is taught in schools. Scientific debates and even scientific knowledge itself can also mirror political debates and political convictions, as Shapin and Schaffer (1989) argued in their classic *Leviathan and the Air-Pump*. But, in so far as it is a power struggle, science is not only potentially political: *science is politics*. Not in the sense of party politics or political/ideological convictions, but in the sense of Michel Foucault's analysis of power/knowledge relations. Scientific knowledge and power cannot be without each other. Nonetheless, I do not contend that this picture of science as a power struggle applies equally well in every case, most likely not. Its explanatory value and its political value can only be judged by each case, by a concrete analysis of a specific instance of scientific knowledge or scientific practice. Foucault did not offer a general theory of scientific knowledge and power, neither is it my intention to do that here.

It may seem that a project, which wants to help us resist science, is undermining science; it may seem decidedly anti-science. This is surely not what we need in times of climate change denial, vaccine refusal and the anti-intellectualism of right-wing populists. As a type of science criticism, however, this project is no more anti-science than literary criticism is anti-literacy. The sort of analysis I am proposing takes science very seriously. Any resistance to science is based on a careful attention to scientific facts and scientific practice. The idea is neither to reject science nor to put its credibility on par with any other opinion, but to gain better understanding of how science works in particular instances when it affects what we can be, what we can do, think and say.

My discussion of the politics of truth is based on the work of Michel Foucault. The first two chapters are directly about Foucault and in the other two chapters Foucault is constantly in the background. Although I will occasionally refer to secondary sources and to lesser extent Foucault's posthumously published work, I am primarily concerned with Foucault's own texts, the work that he prepared for publication during his lifetime. This is the work that best reflects Foucault's considered views at a given time.

Foucault's thought went through a number of transformations. Its development is characterized by Foucault's readiness to change his views, positions, methods, terminology and topics. Foucault was well aware of all this. By the end of the introduction to his *Archaeology of Knowledge*, Foucault writes: "Do not ask who I am and do not ask me to remain the same: leave it to our bureaucrats and our police to see that our papers are in order" (Foucault 1972, p. 17). For Foucault it is even one purpose of writing a book, to transform oneself, to change. In an interview, conducted at the end of 1978 by Duccio Trombadori, Foucault says:

> I write a book only because I still don't exactly know what to think about this thing I want so much to think about, so that the book transforms me and transforms what I think. Each book transforms what I was thinking when I was finishing the previous book. I am an experimenter and not a theorist. I call a theorist someone who constructs a general system, either deductive or analytical, and applies it to different fields in a uniform way. That isn't my case. I'm an experimenter in the sense that I write in order to change myself and in order to not think the same thing as before. (Foucault 2002 [1980], pp. 239–240).

An exposition of Foucault's thought, even a very brief one, would therefore do well to proceed chronologically, locating Foucault's positions and views on the map of his changing thought.

The development of Foucault's thought is commonly divided into three periods, although there are disagreements about the differences between and within these periods and whether there is a single approach, theme or concern that goes through all his work, such as power, critique, or problematization (Koopman 2013). The first period is that of Foucault's *archaeology*, starting with the publication in 1961 of *Folie et déraison*, which was published in an abridged English translation as *Madness and Civilization*,[1] and ending in 1969 with the publication of *L'archéologie du savoir*, which was published in English as *The Archaeology of Knowledge*. The other two major works published in this period are *Naissance de la clinique* (*The Birth of the Clinic*), published in 1963, and the bestseller *Les mots et les choses* (*The Order of Things*), published in 1966.

The second period, spanning the greater part of the seventies, is the *genealogical* period. It is not only characterized by a shift in method, terminology and style, but even more by a new concern: power. The two major works of this period are *Surveiller et punir* (*Discipline and Punish*), published in 1975, and *Histoire de la sexualité I: La volonté de savoir* (*The History of Sexuality I: An Introduction*), published in 1976. Also important are articles and interviews from 1970 to 1977, some of which have

---

[1] Foucault published a revised edition of *Folie et déraison* in 1972 using the title *Histoire de la folie à l'âge classique*, which was the subtitle of the first edition. An unabridged translation of the revised edition was published in 2006 as *History of Madness*, but I will refer here to the work as *Madness and Civilization*. For a further discussion of the revised edition of *Folie et déraison* and its English translation, see Beaulieu and Fillion (2008).

been published in English under the title *Power/Knowledge*. After Foucault's death, transcriptions of his lectures at the Collège de France have been published. They provide important background to the works he published during the 1970s and until his death.

The third and last period covers the last seven years of Foucault's life, 1977–1984, it is sometimes referred to as the *ethical* period. In this period Foucault wrote two more volumes of *The History of Sexuality*, *L'usage des plaisirs* (*The Use of Pleasure*) and *Le souci de soi* (*The Care of the Self*), both of which were published in 1984. Foucault also wrote a fourth volume on sexuality in the 16th century, which was recently published in France as *Histoire de la sexualité 4: Les aveux de la chair* (Foucault 2018; Elden 2018). This period is characterized by Foucault's concern with the subject, or the individual human being, as the creator of his or her own life. It was a Nietzschean concern with an aesthetic sort of ethics: how to act so as to maximize the beauty of one's life and existence. Foucault referred to it as "techniques of the self." A related concern, which is already present in the second volume of *The History of Sexuality*, was with the subject's relation to truth and truth-telling (*parrhesia*) (Foucault 1985 [1983]). Yet another concern during this period is with Kant and the Enlightenment. It was not entirely a new concern. In 1961 Foucault defended, in addition to his main dissertation (*Folie et déraison*), his "thèse complémentaire" for his doctorate degree at the University of Paris. The thesis was an introduction to and translation with notes of Kant's *Anthropology* (*Anthropologie in pragmatischer Hinsicht*). From that time on until the late seventies Foucault refers rarely to Kant. Foucault was nonetheless deeply affected by Kant and, as has been noted by many of his more careful readers, is closer to Kant than most people, especially his critics, may realize.[2]

In the following chapters, I will briefly discuss the central aspects of two of the periods outlined above: first Foucault's archaeology and then his power/knowledge analyses. I conclude my discussion of Foucault with a description of the critical analysis which I am calling "politics of truth." The second half of this book consists of two chapters which are concerned with cases where people become subjects of science. The first of the two concerns left-handers and the second concerns Icelanders. In both cases the subjects have found themselves in a power struggle. In both cases some individual subjects reacted to the effects of power that the sciences were having and in both cases, I argue, they did not have a good idea about how to resist. I propose a way to analyse the struggle and a way for the subjects of science to resist. In these case studies I critically analyse the scientific discourses through which the two groups become subjects of science and discuss the effects of power the discourses have on their subjects.

Foucault sometimes described his work as a toolbox (see Eribon 1991, pp. 125, 237; Foucault 1994 [1974], p. 523). Instead of devising a grand strategy to fight

[2]Han (2002); Djaballah (2008) both explore in very different ways Foucault's relation to Kant; Koopman (2010) makes a strong case for the importance of Kant for Foucault; see also Hacking (1986); Hacking (1995), p. 295 n.2; Rajchman (1985), pp. 105–108; Lee (1997), pp. 93–98; and the discussion of Foucault as a Kantian in Koopman (2013), pp. 13–16.

repression and domination, instead of telling people what to do, he offered some intellectual tools that could be of use in "local battles". Politics of truth consists in such local battles, battles which involve the complex and pervasive dynamics of power and knowledge. By "politics" I do not mean party politics or merely the public sphere, where power struggles are more or less in the open. "Politics" should here rather be understood as any instance of governing and subjugation, any instance of struggle for power—or against power, even when, and especially when, the individuals involved are not aware of a power struggle taking place or them being governed. Modern techniques of power and subjugation are subtle, quiet, pervasive.

In my discussion of Foucault I do not intend to offer a comprehensive account of his intellectual development, or a summary of his works and methods. I merely want to give a brief outline of some of the fundamental aspects of Foucault's thought, in order to give context to and clarify the idea of a politics of truth as a critical analysis.

References

Beaulieu, Alain, and Réal Fillion. 2008. Michel Foucault, *History of Madness*, translated by Jonathan Murphy and Jean Khalfa (London/New York: Routledge, 2006). *Foucault Studies* 5: 74–89.

Djaballah, Marc. 2008. *Kant, Foucault, and Forms of Experience*. New York: Routledge.

Elden, Stuart. 2018. Review: Foucault's Confessions of the Flesh. *Theory, Culture and Society*. https://www.theoryculturesociety.org/review-foucaults-confessions-flesh/. Accessed 15 September 2018.

Eribon, Didier. 1991. *Michel Foucault*. Cambridge, MA: Harvard University Press.

Foucault, Michel. 1972. *The Archaeology of Knowledge, and the Discourse on Language*. New York: Pantheon. Originally *L'archéologie du savoir* (Paris: Gallimard, 1969).

Foucault, Michel. 1985 [1983]. *Discourse and Truth: The Problematization of Parrhesia*. Six lectures given at the University of California at Berkeley, October–November 1983, ed. Joseph Pearson. http://foucault.info/documents/parrhesia/. Accessed 14 September 2018.

Foucault, Michel. 1994 [1974]. Prisons et asiles dans le mécanisme du pouvoir. In *Dits et Ecrits*, vol. II. Paris: Gallimard, 521–525.

Foucault, Michel. 1997 [1978]. What is Critique? In *The Politics of Truth*, ed. Sylvère Lotringer. New York: Semiotext(e). This is a lecture which Foucault gave to the French Society of Philosophy on 27 May 1978.

Foucault, Michel. 2002 [1980]. Interview with Michel Foucault. In *Essential Works, Volume 3: Power*, ed. James Faubion and Paul Rabinow, 239–97. London: Penguin Books. This interview was conducted by D. Trombadori near the end of 1978 and first published in Italian in 1980. An earlier English translation was published in 1991 as *Remarks on Marx: Conversations with Duccio Trombadori*. New York: Semiotext(e).

Foucault, Michel. 2018. *Histoire de la sexualité 4: Les aveux de la chair*, ed. Frédéric Gros. Paris: Gallimard.

Hacking, Ian. 1986. Self-Improvement. In *Foucault: A Critical Reader*, ed. David Couzens Hoy, 235–240. Oxford and Cambridge, MA: Basil Blackwell.

Hacking, Ian. 1995. *Rewriting the Soul*. Princeton, NJ: Princeton University Press.

Han, Béatrice. 2002. *Foucault's Critical Project: Between the Transcendental and the Historical*. Trans. E. Pile. Stanford, CA: Stanford University Press.

Koopman, Colin. 2010. Historical Critique or Transcendental Critique in Foucault: Two Kantian Lineages. *Foucault Studies* 8: 100–121.

Chapter 2
Foucault's Archaeology of Knowledge

> One ought to read everything, study everything. In other words, one must have at one's
> disposal the general archive of a period at a given moment. And archaeology is, in a strict
> sense, the science of this archive. (Foucault 2000 [1966], p. 263)

The main works of Foucault published during the 1960s are "archaeological," in the sense Foucault gave to this term. I shall in this chapter discuss first the origin of the term "archaeology" in his writings and a concept that became central to his mature archaeological work, the *episteme*. In the second section, I shall discuss archaeology as Foucault practiced it in his *The Order of Things*, followed in the third section by a brief consideration of the other archaeological works. This approach puts *The Order of Things* at the centre of Foucault's archaeological work, rather than the theoretical framework for archaeology presented subsequently in *The Archaeology of Knowledge*. In my view, the former offers a more useful view of Foucault's archaeology, it is rich in historical detail and it demonstrates through practice one way of analysing and understanding scientific knowledge and scientific practice. The latter, on the other hand, is abstract and peppered with gratuitously inventive terminology, which has given rise to a plethora of works using Foucault's terms and theoretical constructs in various ways with little attention to historical details and facts. My lack of appreciation for *The Archaeology of Knowledge* is certainly not shared by all Foucault scholars, but I hope my discussion in this chapter further clarifies my reasons for favouring *The Order of Things*.

2.1 The Origin and Development of Archaeology

Foucault's mature archaeological work is about finding historical conditions for scientific knowledge (systems of thought) and explaining sudden changes (historical ruptures) in the organization of scientific knowledge in terms of changes in those historical conditions. Archaeology is thus in some sense Kantian, although it investigates

© The Author(s) 2018
G. Árnason, *Foucault and the Human Subject of Science*, SpringerBriefs in Ethics,
https://doi.org/10.1007/978-3-030-02813-8_2

historical rather than transcendental conditions of knowledge (Koopman 2010, 2013; McQuillan 2010). Foucault's archaeology develops through *Madness and Civilization* and *The Birth of the Clinic*, becomes explicit in *The Order of Things* and is finally presented in its problematic theoretical purity in *The Archaeology of Knowledge*.

In this chapter I am primarily concerned with the works just mentioned, but Foucault did use the word "archaeology" before the publication of *Madness and Civilization*. The first occurrence of the term is already in his first publication in 1954, *Maladie mentale et personnalité*, a book commissioned by Louis Althusser. It was republished in 1962 as *Maladie mentale et psychologie*, with an entirely different second part and conclusion. This later version was published in English translation as *Mental Illness and Psychology* in 1976.[1] This first occurrence of "archaeology" in Foucault's published work refers to "archaic stages" through which the individual evolved according to psychoanalysis (Eribon 1991, pp. 68–69). This suggests a psychological sense of "archaeology", which is barely related at all to the meaning that Foucault gave to the term in his following works.

On the back of my copy of Foucault's *Madness and Civilization* it says "Michel Foucault examines the archaeology of madness." Unfortunately, there is no index to this greatly abridged English translation of *Folie et déraison*, but I have only found two occurrences of the term "archaeology" in the book. The first is in the Introduction: "The language of psychiatry, which is a monologue of reason *about* madness, has been established only on the basis of such a silence. I have not tried to write the history of that language, but rather the archaeology of that silence" (Foucault 1965, pp. x–xi). As Alan Sheridan notes, this term seems to be here "thrown off, almost in passing, as if Foucault is looking around for a word to distinguish what he is doing from 'history'" (Sheridan 1980, p. 14). Foucault says that much in a 1969 interview:

> I first used the word somewhat blindly, in order to designate a form of analysis that wouldn't at all be a history (in the sense that one recounts the history of inventions or of ideas) and that wouldn't be an epistemology either, that is to say, the internal analysis of the structure of a science. This other thing I have therefore called 'archaeology'. (Foucault 1989b [1969], p. 45)

The second occurrence of "archaeology" in *Madness and Civilization* is in the main text, where Foucault describes the classical view of madness as "reason dazzled," *raison éblouie*. This dazzlement of reason "is night in broad daylight" ("c'est la nuit en plein jour") (Foucault 1961, p. 262); and Foucault then elaborates on that with a discussion of the perception of day and night in the classical tragedies:

> Madness designates the equinox between the vanity of night's hallucinations and the non-being of light's judgments. And this much, which the archaeology of knowledge has been able to teach us bit by bit, was already offered to us in a simple tragic fulguration, in the last words of *Andromaque*. (Foucault 1965, p. 111)

What follows is an explication of the last act of Euripides' tragedy *Andromache*, "the last of the great tragic incarnations of madness," (Foucault 1965, p. 112) but there is

---

[1]There is a good reason not to count this book among Foucault's main works: Foucault renounced both editions and tried to prevent the translation of the later version into English (see Eribon 1991, p. 70).

no explanation of the role of the archaeology of knowledge. Again the word seems to be "thrown off, almost in passing."

As far as I can tell, Foucault did not use the term "archaeology" again in his essays or interviews before he wrote his second major work, *The Birth of the Clinic*, and then only in the subtitle: *An Archaeology of Medical Perception ("une archéologie du regard médical")*. It is first in *The Order of Things* (subtitled *An Archaeology of the Human Sciences*, or, in the original, *"une archéologie des sciences humaines"*) that Foucault uses the term "archaeology" in his texts to refer to his own method. In contrast to the absence of the term in Foucault's first two major works, it is quite liberally used in *The Order of Things*. Here it is clear that Foucault's archaeology has taken a specific shape, distinguished not only from history and epistemology, but also from intellectual divisions such as Marxism, phenomenology and, in particular, structuralism. Foucault's relation to all three is complex, except in that he eventually rejected all of them.

The first occurrence of the term "archaeology" in *The Order of Things* (excluding the preface) is informative. Foucault is discussing the ideas of microcosm and macrocosm in the 16th century, first on the superficial level of what he calls opinion (which is the surface knowledge, the subject-matter of history of ideas), then on an altogether different level: "If, on the other hand, one investigates sixteenth-century knowledge at its archaeological level—that is, at the level of what made it possible—then the relations of macrocosm and microcosm appear as a mere surface effect" (Foucault 1971, p. 31). Archaeology has to do with, to use a Kantian sounding phrase, what makes knowledge possible. At its core lies a fundamental distinction between particular bits of knowledge, or surface knowledge, and the background knowledge, or the underlying system of thought, which makes the surface knowledge possible. Within a scientific discipline, rival theories must be "surface effects" of the same underlying structure in order to conflict at all. But Foucault is not so much concerned with the conflict of theories as the transformation of disciplines, or, in other words, ruptures in the history of knowledge and the apparent discontinuity of science. This concern is much influenced by French philosophers and historians of science, in particular Gaston Bachelard and Georges Canguilhem.[2]

To study *archaeology*, in Foucault's sense, is to study the *archive* or the written texts of a period, as he says in the quote opening this chapter. But *archaeology* also refers to the origin of knowledge, as the Greek *arché* (beginning, origin) indicates. This origin of knowledge is not in the objects themselves, nor in the human mind or soul, but in the very conditions for knowledge. These conditions for knowledge are rules or principles within the system of thought, which organize and structure the field of knowledge within a discipline in a given period (and are hence historical).

In *The Order of Things* Foucault describes a system of thought underlying the three disciplines that study life, language and labour. He shows how transforma-

---

[2]For a good introductory discussion of Bachelard and Canguilhem, see Gutting (1989), pp. 9–54; and Dews (1992). Mary Tiles's (1984) book on Bachelard is intended to make his work more accessible to analytic philosophers of science. A critical view of the influence of Bachelard and Canguilhem on Foucault is offered by Ross (2018).

tions of the system of thought manifest themselves in transformations of the respective disciplines, as well as in literature and philosophy. Foucault sometimes called this underlying structure the *positive unconscious* of knowledge (see for example Foucault 1971, p. xi). The underlying structure, the shared background knowledge, is called "unconscious" because no one may be aware of it at the time, and "positive" because it is not a source of error, it does not throw the scientist off the track to knowledge, but is necessary for there to be knowledge at all. Foucault acknowledges the Kantian aspect of his view, and referred occasionally to the underlying structure as the "historical a priori" of knowledge (see for example Foucault 1971, p. xxii; and Foucault 1972, Chap. 5). Foucault used the word *connaissance* to refer to the surface knowledge and *savoir*, or *episteme* (from the Ancient Greek meaning knowledge), to refer to the underlying structure, or the depth knowledge.[3] Here I will use Foucault's term *episteme* to refer to the underlying system of thought.

Until the very end of his archaeological period, Foucault appears to be quite undecided about the scope of the *episteme*. Throughout *The Order of Things* the *episteme* appears to be an underlying structure common to the disciplines under investigation as well as to philosophy, literature and art. Thus the *episteme* appears to be 'global' in its scope. In the 'Foreword to the English edition' of *The Order of Things*, Foucault claims retrospectively that his study "was to be not an analysis of Classicism in general, nor a search for a *Weltanschauung*, but a strictly 'regional' study." And he adds in a footnote: "I sometimes use terms like 'thought' or 'Classical science', but they refer practically always to the particular discipline under consideration" (Foucault 1971, p. x).

If the archaeological level is specific to particular disciplines, and each discipline has its own *episteme*, then there must be something that differentiates them. But Foucault's comparison of disciplines in *The Order of Things* appears to present only what they have in common at the archaeological level, indicating that he considered the *episteme* to be more 'global' at the time of writing than he thought in retrospect.

Following the lines just quoted, Foucault claims he wished to "describe [...] an epistemological space *specific to a particular period*" (emphasis added). And only a few lines below that he writes: "Taking as an example the period covered in this book [*The Order of Things*], I have tried to determine the basis or archaeological system common to a whole series of scientific 'representations' or 'products' dispersed throughout the natural history, economics, and philosophy of the Classical period" (Foucault 1971, pp. xi–xii).

When one reads *The Order of Things* one can't help considering the *episteme* to belong to a culture at a certain period, rather than just to a single discipline. But Foucault certainly had some second thoughts about the scope of the *episteme*, and in *The Archaeology of Knowledge* he strictly limits its scope to the particular scientific discipline (or, more generally, to the "discursive formation"). His concern at that point was that a 'global' interpretation of the *episteme* would invite the cultural totalizations inherent in histories that are centered on *Zeitgeist* or worldviews

---

[3]The terms "surface knowledge" and "depth knowledge" are borrowed from Ian Hacking, see Hacking (1979), p. 42.

(Foucault 1972, p. 16). In the end, however, Foucault acknowledges the complexity of the relation between the *episteme* and the scientific disciplines of a given period: some archaeological rules are specific to a single discipline, others are common to two or more disciplines and some may even be characteristic of the general thought of a period. What might be considered his final word on the scope of the archaeological rules is the following quote from a pamphlet Foucault prepared in 1969 during his campaign for a chair at the Collège de France, describing his research and publications: "Certain of these rules are specific to a single domain; others are shared by several. It is possible that, for a particular period, there are others that are generalized" (Quoted in Eribon 1991, p. 216).

Whatever its scope may be, the *episteme* determines what the scientist *can* say within his discipline (and not what he *does* say). It does not determine what is true or false, but what *can* be true or false. Foucault says he tried to

> explore scientific discourse not from the point of view of the individuals who are speaking, nor from the point of view of the formal structures of what they were saying, but from the point of view of the rules that come into play in the very existence of such discourse: what conditions did Linnaeus (or Petty, or Arnauld) have to fulfill, not to make his discourse coherent and true in general, but to give it, at the time when it was written and accepted, value and practical application as scientific discourse—or, more exactly, as naturalist, economic, or grammatical discourse? (Foucault 1971, p. xiv)

In *The Order of Things*, archaeology's domain of inquiry is *scientific discourse* as it appears in written texts.[4] Archaeology looks for the rules that make the appearance of those written sentences possible. The rules that function on the archaeological level are neither grammatical nor logical; they are perhaps most usefully described with a phrase I mentioned above: as "the historical a priori" of a discourse. The rules are historical simply in the sense that they may differ from one period to another, and they may be called a priori in the Kantian sense that they make particular kinds of knowledge possible. But they are known empirically and not independently of experience by means of, say, a transcendental deduction, and are therefore strictly speaking not known a priori. Foucault seems to have chosen the label "historical a priori" for the reasons given above: the rules are the historical conditions for knowledge.

It may appear self-contradictory to speak of a "historical a priori." But just as a definition of a unit of length is necessary for the practice of measuring length, some rules may be necessary for there to be knowledge. And those rules may change over time, just as definitions of a unit of length may do. As such definitions are stipulations, they are known a priori (it would not make sense to measure the defining unit to make sure it is of the right length). The archaeological rules may be thought of as a priori in a similar manner; they are not stipulations but in so far as they make certain knowledge

---

[4]I am using the terms "discourse" and "text" in a non-technical sense as Foucault did in *The Order of Things*. In *The Archaeology of Knowledge* a discourse denotes both a group of statements and a linguistic practice (Foucault 1972, pp. 80, 107), in either case discourse is strictly a linguistic entity. Social relations, institutions and economic processes are nondiscursive factors that have complex relations to discursive factors, relations that are not made all that clear in Foucault's archaeological work (Gutting 1989, pp. 256–258; but see also Tiisala 2015, pp. 666–668).

possible, it would be pointless to justify them empirically in terms of that knowledge. In both cases the necessity in question is not the logical or transcendental necessity of the *particular* definition and *particular* archaeological rules, but the practical necessity that *some* definition and *some* archaeological rules be in place to make scientific practice possible.

Foucault did not, however, take this Kantian description any further than using it as a label, because, to repeat, the rules are known through empirical, historical study, not independently of experience, and hence not really known a priori in the usual sense of the term. Foucault's use of the term 'a priori' should not be understood strictly in a Kantian sense, there can be no transcendental deduction of historical categories. Foucault explicitly rejects "formal a prioris" and subjecting the history of thought to transcendence; he wants to present discourse without these categories which he believes will prejudge accounts of the development of discourses by forcing one to look for their rational development towards a given goal (teleologies) and for their pure origins and timeless truth (Foucault 1972, p. 203). Foucault's aim is to give a systematic description of discourses in their temporality and with a minimum of philosophical prejudice; and in that sense archaeology is strictly empirical.

Foucault's rejection of origins, teleologies and the rational purity of knowledge bears not only the obvious marks of Nietzsche's influence, but also that of French historians mentioned above: Bachelard and Canguilhem. In the introduction to *The Archaeology of Knowledge* Foucault refers to Bachelard's ruptures in the history of knowledge and their upsetting any search for origins. And he refers to Canguilhem's analyses of concepts, which he takes to show that

> the history of a concept is not wholly and entirely that of its progressive refinement, its continuously increasing rationality, its abstraction gradient, but that of its various fields of constitution and validity, that of its successive rules of use, that of the many theoretical contexts in which it developed and matured. (Foucault 1972, p. 4)

The *episteme* is manifested in the texts scrutinized by archaeology. These texts are to be found not only in learned treatises and journals, but also in literature, in leaflets and government orders, in the plans and written rules of institutions and so forth. The archaeologist is not much concerned with individuals; he considers the whole field of texts related to a discipline of a given period—he must read everything—more or less regardless of who produced the texts, what they thought or what influenced them. Foucault was adamant to exclude phenomenology and psychology from his archaeological method. By "freeing" discourse from these disciplines (as Foucault often put it), the facts of discourse could be presented without any demands or requirements to give account of the intentions, thoughts and acts of the individual subjects of discourse. This allows discourse to be analyzed in completely different terms and by different unities, namely those of relations between statements, relations between groups of statements, and relations between statements, groups of statements and technical, economic, social and political events (Foucault 1972, pp. 29, 203; and more generally ibid., Part II).

Archaeology is not of a highly abstract nature. It is concerned with actual texts and their relation to institutions, social practices and knowledge. Archaeology is badly served when it is discussed on a purely theoretical level, to which Foucault's own *Archaeology of Knowledge* bears witness. One of the main virtues of Foucault's archaeology is its concern with fact and historical detail that is not pursued merely for the sake of fact and detail, but for the sake of its philosophical and political purposes.

It has been said that Foucault could have done better when it came to historical facts. He has been accused of making a very selective use of historical records, misrepresenting historical facts and over-generalizing, in particular in *The Order of Things* (see for example Gutting 1989, pp. 175–179). As Hacking has argued, Foucault can be forgiven occasional historical inaccuracies because his aim is not so much to present an accurate and comprehensive history as to practice a kind of politics and a kind of philosophy:

> Now not only do we find that the facts are sometimes not quite right, that they are overgen-eralized, and that they are squeezed into a model of brusque transformations; we also find that many of Foucault's dramas have already been told in calmer terms, by other people. No matter. His histories stick in the mind. We can add our own corrective footnotes at leisure. These histories matter because they are in part political statements. They are also what I call philosophy: a way of analyzing and coming to understand the conditions of possibility for ideas. (Hacking 1986 [1981], p. 30)

The best way to get to know archaeology is to see it in action and Foucault's clearest example of archaeology in action is *The Order of Things*. Let me then in the next section take a closer look at archaeology in *The Order of Things*, and then, in the section that follows, briefly consider archaeology in *Madness and Civilization* and *The Birth of the Clinic*, as well as Foucault's theoretical discussion in *The Archaeology of Knowledge*.

2.2 Archaeology in Action

In *The Order of Things*, Foucault's archaeological inquiry finds two breaks in the archaeological system, the *episteme*, of Western culture (Foucault 1971, p. xxii). The first is around the middle of the 17th century, around the time of Descartes. It marks the end of the Renaissance and the beginning of what is known in France as the "Classical age." The second break is at the beginning of the 19th century, the Modern age, following the French revolution. Both of the "breaks" take place at a time that is perhaps more a turning point in French history than in the history of other Western cultures (Hacking 1979, p. 46). In *The Order of Things* Foucault saw signs of the third break already taking place (see also Hacking 1986 [1981], p. 31).

The bulk of *The Order of Things* is about the three realms of knowledge concerned with language, life and labor, during the Classical age: general grammar, natural history and analysis of wealth, respectively. It is preceded by a discussion of Renaissance knowledge, and followed by a discussion of the human sciences that took the place of general grammar, natural history and analysis of wealth in the Modern age, namely philology, biology and political economics. Foucault emphasizes that the relation between the three sciences in the Classical age on one hand and the three in the

Modern age on the other, is not that of gradual development or progress. The latter replaced the former; they are divided by a historical discontinuity, a break.

Four related and overlapping areas of Foucault's analysis of an *episteme* can be identified:

(1) Rules and concepts that organize or structure knowledge;
(2) the role of language;
(3) semiotics (the nature of signs); and finally
(4) hermeneutics (how signs are interpreted).

The first is most fundamental to the *episteme*, largely determining the other three, which nonetheless belong to the *episteme* rather than the surface knowledge. These four areas can be identified in Foucault's discussion of the Renaissance *episteme* as follows. In the first area (organizing rules): *Resemblance* plays a key role in the organization of Renaissance knowledge. In the second (language): Language is not only the product of speakers and writers, it constitutes the world. Texts are not only in books; the whole world is a text. The third (semiotics): Signs are everywhere; everything is a sign signifying whatever resembles it. The system of signs "contained the significant, the signified and their 'conjuncture'." This conjuncture is the resemblance of the significant to the signified, hence "the three distinct elements of this articulation are resolved into a single form" (Foucault 1971, p. 42). And finally the fourth area (hermeneutics): All interpretation of signs is guided by resemblance. In Renaissance knowledge the two last areas are "superimposed": a sign signifies whatever resembles it, hence the act of interpretation simply consists in bringing out the resemblances of signs (Foucault 1971, p. 17). This results in a picture of language in the 16th century in which language operates on three levels: the signs as marks inscribed on the world, above which is commentary as discussion and interpretation of the signs, and below which is the text as the meaning behind the inscriptions, put there in books by man, in nature by God (Foucault 1971, p. 42; for God as the author of the "book of nature," see for instance ibid., pp. 33, 59.). The interaction between these levels is determined by *resemblance*, in terms of which Renaissance knowledge is organized and constructed (or rather endlessly gathered). A wonderful example of this organization is Aldrovandi's *Historia serpentum et draconum*, cited by Foucault:

> And indeed, when one goes back to take a look at the *Historia serpentum et draconum*, one finds the chapter "On the serpent in general" arranged under the following headings: equivocation (which means the various meanings of the word *serpent*), synonyms and etymologies, differences, form and description, anatomy, nature and habits, temperament, coitus and generation, voice, movements, places, diet, physiognomy, antipathy, sympathy, modes of capture, death and wounds caused by the serpent, modes and signs of poisoning, remedies, epithets, denominations, prodigies and presages, monsters, mythology, gods to which it is dedicated, fables, allegories and mysteries, hieroglyphics, emblems and symbols, proverbs, coinage, miracles, riddles, devices, heraldic signs, historical facts, dreams simulacra and statues, use in human diet, use in medicine, miscellaneous uses. (Foucault 1971, p. 39)

Knowledge of the serpent consisted in everything that had been said, seen or heard about serpents, no matter if it was direct observation, myth, dream or story. Language was found not only on the pages of books, but all over nature and its creatures.

Nature and the word are intertwined, forming "one vast single text" whose truth is "coeval with the institution of God" (Foucault 1971, p. 34). Knowledge consisted in a *commentary* on all those signs, in which they are organized according to resemblance. That way, resemblance constitutes both the form and content of Renaissance knowledge. Resemblance has just enough norms, rules and criteria to make stories about resemblance possible, but not enough to put limits to them. One can tell almost endless stories of resemblances, and that is what happens in Renaissance knowledge: knowledge piles up, but with a bare minimum of rules and norms, and, as everything is like everything else, everything is knowledge and nothing is. Foucault notes that the main characteristic of Renaissance knowledge is that it is "plethoric, yet absolutely poverty-stricken." It is nothing but endless accumulation of resemblances, of similitudes and analogies (Foucault 1971, p. 30).

One fundamental point, which Foucault makes in his discussion of Renaissance knowledge, concerns rationality. The *episteme* is a standard of rationality, so to speak. It determines what thought counts as rational. Two examples: First, Pierre Belon made the first known comparison of the human skeleton and that of birds in the 16th century. Was he an early precursor of Darwin and the theory of evolution? No, it just so happened that he made a comparison of resemblances that would in 19th century be seen as an example of comparative anatomy. In the 16th century Belon's comparison was just like any other collection of resemblances, it "is neither more rational nor more scientific than an observation such as Aldrovandi's comparison of man's baser parts to the fouler parts of the world, to Hell, to the darkness of Hell, to the damned souls who are like the excrement of the Universe" (Foucault 1971, p. 22; Belon 1555, p. 37). It is just by accident, and quite a rare one, that our knowledge coincides at this point with that of the 16th century. In a second example, Aldrovandi gathered knowledge about serpents in the 16th century (see above). Roughly a century later, the great natural historian Buffon could not make head or tail of this "hotch-potch of writing" (Foucault 1971, p. 39). Foucault comments:

> Aldrovandi was neither a better nor a worse observer than Buffon; he was neither more credulous than he, nor less attached to the faithfulness of the observing eye or to the rationality of things. His observation was simply not linked to things in accordance with the same system or by the same arrangement of the *episteme*. For Aldrovandi was contemplating a nature which was, from top to bottom, written. (Foucault 1971, p. 40)

The kind of claims Foucault makes here invite a host of problems, concerning rationality, relativism and more specifically incommensurability, that have been widely discussed among Anglo-American philosophers at least since the publication of Thomas Kuhn's *Structure of Scientific Revolutions* in 1962. The implications of Foucault's view for rationality are indeed similar to the implications of Kuhn's *Structure*, but Foucault does not take these issues up as a problematic as Kuhn did and I will not try to speculate how Foucault would or could have dealt with them.

In the Classical age *representation* replaced resemblance as an organizing concept. Language became transparent, instead of being a complex web of words and things it now simply represents the world (Foucault 1971, p. 43). The new "categories" for scientific knowledge are *difference, identity, measurement* and *order* (I shall discuss them below). Foucault discusses three figures who manifest, in different ways, the transformation of the Western *episteme*. In his *Novum Organum*, Francis Bacon criticizes Renaissance knowledge from within, warning that "the human Intellect ... feigns parallels, correspondents, and relations that have no existence." Foucault claims that here it is "sixteenth-century thought becoming troubled as it contemplates itself and beginning to jettison its most familiar forms." Then we have Descartes after the transformation, laying down the rules for proper investigations of nature in his *Rules for the Direction of the Mind* (Foucault 1971, pp. 52–56). The third figure is no less interesting than the two great thinkers, it is Don Quixote, doomed to wander forever in the no man's land between the Renaissance and Classical *epistemes*.

With the emergence of the Classical *episteme*, a whole new ordering of the knowledge of nature came about, a much more structured order whose clearest manifestation is the *ordered table*: "Despite their differences, these three domains [natural history, analysis of wealth and general grammar] existed in the Classical age only in so far as the fundamental area of the ordered table was established" (Foucault 1971, p. 73).[5] Knowledge is now put into timeless, ordered tables, such as Linnaeus' tables in his *Systema naturae* (1735), rather than being endlessly collected and paraded.

In the story as "the history of ideas" tells it, Descartes and Newton are central figures in shaping the sciences of the Classical age. Here "mechanics" and "mathematics" are key terms: the breakthrough of the new science of Descartes and Newton consists in representing nature as mechanical and mathematical. Foucault's archaeology finds a more general characteristic of the age, which is then a fundamental element of the Classical *episteme*. This element is neither a mechanical view of nature nor the attempt to mathematicize nature. Foucault calls it *mathesis*, "understood as a universal science of measurement and order" (Foucault 1971, p. 56). In French, the word mathesis is almost unknown outside of Foucault's *The Order of Things*, but it has a long history in English, meaning, according to the *Oxford English Dictionary*, "mental discipline; learning or science, esp. mathematical science."[6] By "universal science of measurement and order," Foucault does not mean that there was such a scientific discipline (there wasn't), but rather that the study of measurement and order became an essential part of all scientific work. It is the organization of knowledge according to rules of measurement and order that gives rise to mechanical and mathematical representations of nature, as well as other surface effects of the Classical *episteme*. Descartes and Newton are not the originators or main force of this transformation, they and their work are its manifestations.

In the Classical age resemblance would not be accepted "until its identity and the series of its differences have been discovered by means of measurement with a common unit, or, more radically, by its position in an order (Foucault 1971, p. 55). "Measurement" and "order" are two ways of comparing things. Objects are compared in terms of "measurement," with reference to an external standard or unit; or in terms of "order," involving only the objects ordered. That is to say, "measure-

---

[5] Here Foucault applies the episteme to a whole period; see Foucault (1971), pp. 50–77, for a general discussion about "the episteme of the Classical age"; see also Gutting (1989), p. 155.

[6] I owe this observation to Ian Hacking, personal communication.

ment" involves comparing things with a standard or unit that is external to the objects compared, while "order" involves comparing things according only to characteristics of the objects themselves rather than anything external to them. The relation of the four "categories" mentioned above, difference, identity, measurement and order, is roughly that the comparison by *measurement* and *order* involves establishing the *identity* and *differences* of the things compared. Comparison is here based on discrimination, rather than the endless gathering of similitudes in 16th-century knowledge. Foucault takes a number of examples, all from Descartes, to illustrate this new thinking. The new categories, in particular their combination in *mathesis* as a universal science of measurement and order, along with language withdrawing from the world and becoming a transparent medium instead of a source of knowledge, are a much deeper and more general phenomenon than "the particular fortunes of Cartesianism" (Foucault 1971, p. 56).

Above I identified four areas within Foucault's analysis of the archaeological level, the *episteme*, of the 16th century. The same areas can be identified when it comes to the Classical age. I have already discussed the first two, the fundamental concepts and configurations of knowledge, and the role of language. Foucault has also a good bit to say about the second two, the nature of signs and methods of interpretation, or semiotics and hermeneutics.

In the Classical age, according to Foucault, signification became for the first time a binary relation between a sign and a signified. Signs were defined with respect to three variables: *certainty*, *type* and *origin*. A sign can be *certain or probable*, as breathing is a certain sign of life or pallor a probable sign of pregnancy (Foucault 1971, p. 58). This makes the sign thoroughly epistemic; it is not a part of nature anymore; it functions only in relation to knowledge. It is these changes in the system of signs that make the modern concept of probability possible (this has been further investigated in Hacking 1975). Impressions are linked to each other as signs are linked to what they signify, so given one impression there can now be a certain probability of another.

The *type* variable is somewhat obscure. The sign can either "belong to the whole that it denotes (in the sense that a healthy appearance is part of the health it denotes) or be separate from it (in the sense that the figures of the Old Testament are distant signs of the Incarnation and Redemption)." Foucault adds that the two possibilities are closely related: when the sign is linked for the first time (psychologically) to the signified both must have been parts of the *same* perception. Consequently, the sign and the signified must have been analyzed as *distinct*, and only then could the sign be linked to the signified (Foucault 1971, pp. 58, 60–61).

The third variable is *origin*: Signs could have two kinds of origin, natural or conventional. In the 16th century signs were considered primarily natural, and the conventional signs "owed their power only to their fidelity to natural signs." In the 17th century it is the conventional signs that matter. The conventional sign could be invented in such a way that "it will be simple, easy to remember, applicable to an indefinite number of elements, susceptible of subdivisions within itself and of combination with other signs" (Foucault 1971, pp. 61–62). The sign had to suit the language of calculability.

Foucault remarks that where history of ideas might find in the transformation of the sign system "a tangled network of influences," an archaeological investigation will show that

> the dissociation of the sign and resemblances in the early seventeenth century caused these new forms—probability, analysis, combination, and universal language system—to emerge, not as successive themes engendering one another or driving one another out, but as a single network of necessities. And it was this network that made possible the individuals we term Hobbes, Berkeley, Hume or Condillac. (Foucault 1971, p. 63)

I have noted that in the Classical *episteme* resemblance is replaced by representation as a fundamental organizing concept, but so far I have barely mentioned representation, except that language is supposed to represent the world in a neutral and transparent manner. The sign plays a central role in representation: instead of the sign being connected to the signified through resemblance, signification has become the immediate link of one idea with another, the idea of the thing representing and the idea of the thing represented. The ternary relation of signification by resemblance (the sign, the signified, resemblance) has been replaced by the binary relation of signification by representation (the representing, the represented). Representation does not replace resemblance as the third term in the sign relation; representation is immediate and transparent.

The hermeneutics of the Classical age are joined again with the semiotics, not in resemblance as in the 16th century, but in the "power proper to representation of representing itself" (Foucault 1971, p. 66). The representation of representation can, claims Foucault, be observed in "Las Meninas," painted by Diego de Silva y Velázquez in 1656—Foucault's analysis of that painting forms the first chapter of *The Order of Things*. In this painting the painter can be seen painting a couple, but from the viewpoint of the couple, who can be seen in a mirror on the wall. It is not immediately obvious what the possibility of double representation has to do with interpretation. Foucault has told us that in the Classical age the meaning of signs is immediate (representation is transparent), i.e., the meaning of a sign is immediately given once the sign is given. Interpretation, then, is merely representation of representation, that is, the mere repetition of what the sign has already said.

Foucault describes the "configuration" of Classical knowledge, the archaeological level that determines what claims can be true-or-false in that period, in a number of slightly different ways. At one point he characterizes the Classical *episteme* as having four elements, which in a quick summary are *analysis* of measurement and order instead of a hierarchy of analogies in Renaissance knowledge, possibility of complete *enumeration* instead of infinite interplay of similitudes, *certainty* through enumeration instead of constantly changing similitudes, and *discrimination* instead of drawing things together (Foucault 1971, p. 55).

At another point the fundamental element of the Classical *episteme* is the *mathesis* mentioned above, characterized by: (1) the relation between things being thought of in terms of order and measurement, or even just order as measurement could be reduced to order; and (2) new fields of study being based on *mathesis* (as a science of order) rather than mathematization or mechanism. This science of order, or *mathesis*, required analysis as a method and its instrument was not algebra but the new system of signs (Foucault 1971, p. 57).

At a third point, Foucault characterizes the Classical *episteme* in terms of three elements: *mathesis*, *taxinomia*, and *genetive analysis*. Here the *mathesis* mentioned above has been split in two, *mathesis* dealing with ordering "simple natures" using algebra and *taxinomia* dealing with ordering "complex natures" using the system of signs. Genetive analysis is like *taxinomia*, except that while *taxinomia* establishes tables of visible differences (spatial order of things), genetive analysis sets up chronological tables or progressive series (temporal order of things) (Foucault 1971, p. 57).

The three formulations give less importance to the theory of signs than it deserves, considering Foucault's lengthy discussion of signs, and neither representation, which is supposed to be *the* fundamental concept or organizing principle replacing resemblance, nor the transparency of language are mentioned. The reason for all three is, I think, that the formulations give us certain details of a bigger picture and the picture itself is representation. The simple and complex natures that are ordered, organized and that constitute knowledge in the end, are representations, which are in turn signs, and language is their neutral medium. Representation, signs and language are the picture; mathesis, taxinomia and genetive analysis are details of its structure.

So far I have summarized and offered some comments on Foucault's general discussion in *The Order of Things* of the Renaissance *episteme* and the Classical *episteme*. Foucault's main concern is the detailed study of the archaeology of three domains of knowledge in the Classical age, general grammar, natural history and the analysis of wealth. It is not necessary here to go into the details of that study; we have already seen archaeological analysis in practice while the stage was being set for the study of the particular scientific disciplines. At this point I am more interested in the general archaeological level of knowledge in the three periods (Renaissance, Classical age and Modern age) and the breaks between them. Let me therefore turn to the second break, around 1800, when Western knowledge entered the Modern age.

Foucault calls the principle of the new *episteme* History, with a capital H. It is of course misleading to call "History" *the* organizing principle of the modern *episteme*. As Foucault notes, for "an archaeology of knowledge, this profound breach in the expanse of continuities … cannot be 'explained' or even summed up in a single word" (Foucault 1971, p. 217). The Renaissance *episteme* came close to be sufficiently summed up in the word *resemblance*, although resemblance was further divided in four types which Foucault called convenience, emulation, analogy and sympathy; and the omnipresence of language played a big role obviously not conveyed by the word "resemblance." The Classical *episteme* was summed up in *representation*, but as we saw it was more usefully characterized by the ordered table, a concern with identities and differences of objects of knowledge, as well as by *mathesis* and *taxinomia*—not to speak of changes in the role of language and in the sign system.

So, what about the Modern *episteme*? The keywords here are Analogy, Succession, History and organic structures. These terms are not to be taken literally, they can be seen as labels, or "umbrella terms," grouping together a number of concepts and relations. Instead of ordering identities and differences by means of *mathesis* and *taxinomia*, the Modern *episteme* orders objects of knowledge as organic structures ("internal relations between elements whose totality performs a function") by analogy with their relations and their temporal succession—the emphasis on the last item apparently giving rise to the term History in Foucault's analysis. In the 19th century "knowledge is no longer constituted in the form of a table but in that of a series, of sequential connection, and of development" (Foucault 1971, p. 262).

The most obvious manifestation of the new concern with organic structure is in the emergence of the field of biology, which replaced natural history. Because of the radical break at the archaeological level, Foucault insists that the former replaced the latter, rather than one having grown out of or developed from the other. Here Foucault uncharacteristically credits the genius of a single man for the transformation of thought from the category of taxinomia to that of organic structure. This man was the French natural scientist Jean-Baptiste de Monet de Lamarck (1744–1829), who in his *La Flore française* (1778) made a distinction between two tasks of botany: determining the place (and hence the name) of an individual in an ordered table and "the discovery of the real relations of resemblance, which presupposes an examination of the entire organic structure of species" (Foucault 1971, p. 230).

Foucault does not claim that the concept of organic structure did not exist before the end of the 18th century; it did. His claim is rather that it had never been used before then "for ordering nature, as a means of defining its space or delimiting its forms" (Foucault 1971, p. 231). The new importance of the organic structure also resulted, claims Foucault, in the opposition between the organic and inorganic becoming fundamental, replacing the division of nature into three or four "kingdoms." And this is the fundamental change, not the triumph of vitalism over mechanism, as history of ideas would have it. The fate of vitalism and mechanism, like that of Cartesianism, are mere surface effects of archaeological changes (Foucault 1971, p. 232). Their fate is not fundamental to the transformation of scientific thought, but rather a consequence of changes in the configuration of the *episteme*.

The other two sciences studied by Foucault, those concerned with language and labor, were transformed in a similar manner. The science of language changed more slowly than the others, because the transformation that destroyed general grammar and created philology and comparative grammar was the transformation of representation itself. To give a brief example, it was not the representational character of language that mattered anymore, but its grammatical structure, i.e., the relation between the elements of language. Languages are now compared not in terms of the meanings of their words, but in terms of their structural similarities, not in terms of the words but in terms of the rules which link the words together (Foucault 1971, pp. 232–236).

In the Classical age the analysis of wealth had focused on human needs as the force that turns the wheels of economy; the goods exchanged were "objects of need representing one another." In 1776 Adam Smith, "the father of modern economics," made the concept of labor fundamental to economic theory in his classic *Wealth of the Nations*. Analysts of wealth did discuss labor long before Adam Smith, and Adam Smith was still concerned with the analysis of wealth (as objects of need), but the transformation consists in the exchange of goods becoming measured in labor, a concept which, according to Foucault, cannot be further analyzed as a representation. In other words, the concept of labor changed roles, from being an effect of human needs and the exchange of objects of need (both of which Foucault links with representation), to being the fundamental measure of economic exchange (which has nothing to do with representation). In this way the emergence of modern economics and the destruction of the analysis of wealth is tied to the transformation of the *episteme*, in particular the new and limited role of representation in the organization of knowledge (Foucault 1971, pp. 224–225).

The transformation of the *episteme* had a "second phase." The first phase could be linked with the names of Adam Smith in economics (focus on labor and production), Lamarck (focus on the organic structure) and William Jones or Coeurdoux in philology and comparative grammar (focus on grammatical structure, as in their comparison of Sanskrit with Greek and Latin). The second phase Foucault links with the names of Ricardo in economics, Cuvier in biology, and Bopp in philology and comparative grammar.[7] It is only after this second phase that representation, epitomized in the ordered table, has fully withdrawn and the historical sequence, i.e., History, with a capital H, taking over at the archaeological level of Western knowledge.

So far I have tried to describe Foucault's analysis of the fundamental concepts organizing the Modern *episteme*. Something remains to be said about the changing role of language, semiotics and hermeneutics. During the Classical age, language had been considered the transparent medium of representation. It is only by the end of the 18th century that language loses this neutral representational character. Language becomes dense again, like it was in the 16th century, and it becomes for the first time one object of study among others. Foucault has practically nothing to say about a new semiotics in the Modern age; according to him even the "new" semiology of Saussure refers back to the Classical sign system rather than to a Modern semiotics. The new "thickness" or density of language results, however, in a renewed emphasis on hermeneutics, on *exegesis*:

> The first book of *Das Kapital* is an exegesis of "value"; all Nietzsche is an exegesis of a few Greek words; Freud, the exegesis of all those unspoken phrases that support and at the same time undermine our apparent discourse, our fantasies, our dreams, our bodies. (Foucault 1971, p. 298)

Hermeneutics, as a philosophical school, is itself born in the 19th century, the offspring of philology in general and biblical exegeses in particular. Let's not forget that Nietzsche's academic background was in philology, to which he made some contribution before turning to writing texts of a more philosophical nature.

In the penultimate chapter of *The Order of Things* Foucault makes the rather dramatic claim that not only life, language and labour came to exist as objects of

---

[7]David Ricardo (1772–1823) was a British national economist and advocate of classical national economics; Georges Baron de Cuvier (1769–1832) was a French zoologist and paleontologist, founder of comparative anatomy and the "catastrophe theory" (according to which extinct species were destroyed in natural catastrophes); and Franz Bopp (1791–1867) was a German linguist and the founder of comparative linguistics.

knowledge by the end of the 18th century, but man himself. Foucault devotes the rest of the book to a discussion of this claim and goes on to predict that if the epistemic arrangements that made man possible were to disappear, then "one can certainly wager that man would be erased, like a face drawn in the sand at the edge of the sea" (Foucault 1971, p. 387). Although Foucault had already in *The Birth of the Clinic* made the claim that man is a recent invention (Foucault 1973, p. xiv), it reappears here with little warning, boding the possible demise of the sciences of man. Foucault is of course not denying that man existed, although he rhetorically states that before "the end of the eighteenth century *man* did not exist" (Foucault 1971, p. 308). What is being denied is that man existed as an object of study before the Modern age. In the Classical age man was the knowing subject, but not one among the many objects to be known.

Foucault mentions two objections to his claim: (1) the *human* sciences of the Classical age, general grammar, natural history and the analysis of wealth, were very much concerned with human matters, for them man must exist; and (2) the Classical age accorded great attention to human nature. Foucault's answer to the first objection is that in the Classical age man was dealt with as species or genus, as "the controversy about the problem of races in the eighteenth century testifies to" (Foucault 1971, pp. 308–309). There was no *domain* specified as proper to man. Foucault simply dismisses the second objection saying that "the very concept of human nature, and the way in which it functioned, excluded any possibility of a Classical science of man" (Foucault 1971, pp. 308–309).

Although Foucault's claim that man (as an object of science) has just recently emerged has been called "the thesis" of *The Order of Things* (for instance in the jacket blurb of the English translation), it appears rather as an odd twist in the narrative, it's in a way a surprise ending of the story. His analyses of the underlying "structures," or organizing principles, that make knowledge possible, is really the main plot and the most important lesson to draw from this work. According to that main plot, "man" as an object of study is a mere subplot, a surface effect of the modern *episteme*. This subplot, however, provides one of the main political and polemical points of *The Order of Things*: That the human sciences and their knowledge of man is contingent and might disappear like last year's fashion.

After publishing *The Order of Things*, Foucault attempted to systematize his theoretical framework in his *Archaeology of Knowledge*. It was not a summary of his methodological approaches in his previous works, but rather a further development of the archaeological method. Foucault "struggled violently with notions of enunciation, discursive formation, regularity, and strategy" (Eribon 1991, p. 191), while at the same time experiencing the political turmoil, not of France, but of Tunisia in 1968. He completed the book before returning to France, and it was published in 1969. In his subsequent work he dropped almost all of the archaeological terminology and theory he had constructed as his interests turned to a different subject: Power. Before I turn to Foucault's discussion of knowledge and power, let me discuss briefly his other archaeological writings.

2.3 Madness, Medicine, Theory

Foucault's first major work, *Folie et déraison*, was translated into English and published in an abridged version as *Madness and Civilization* in 1965 (an unabridged translation of the book's revised edition was published in English in 2006 as *History of Madness*). Foucault's history of madness is not a history of mental illness or psychiatry, but a history of how madness and reason were divided and became contraries during the 17th century. It is not a history of madness as a scientific object, but a history of how it happened that madness became a scientific object in the first place, as something to be known and to be treated as illness. In the background of Foucault's history there is a romantic idea of a "pure" madness, which becomes subjugated and silenced by reason and science. Later, for instance in *The Archaeology of Knowledge*, Foucault would deny that there is any pure madness that could be separated from the meanings, values and practices that constitute it, taking a strongly nominalist perspective:

> Mental illness was constituted by all that was said in all the statements that named it, divided it up, described it, explained it, traced its developments, indicated its various correlations, judged it, and possibly gave it speech by articulating, in its name, discourses that were to be taken as its own. (Foucault 1972, p. 32)

In *Madness and Civilization* Foucault identifies the periods and breaks which appear again in *The Birth of the Clinic* and *The Order of Things*. Two events mark the breaks in the history of madness: first the creation of the Hôpital Général in Paris and the "great confinement" of the poor in 1657; and second the liberation of the prisoners of Bicêtre in 1794 (Foucault 1965, pp. xi–xii). The largest part of *Madness* is concerned with what happened between the two dates, that is, how the experience of madness changed during the Classical age. The main characteristics of the Classical experience of madness were the idea of madness as unreason, as the very opposite of reason, and the exclusion of the mad, for instance through confinement and other administrative controls. Foucault attempts to analyse the "deeper level" underlying these characteristics, practicing for the first time something like an archaeology—although he uses neither word, "archaeology" or "episteme." Gary Gutting has identified the following seven steps in Foucault's archaeological analysis of the Classical experience of madness:

(1) a characterization of madness as a phenomenon of *mind-body unity*;
(2) a correlative grounding of madness in *passion*;
(3) a recognition of madness as *delirium*—that is, the taking of the unreal for the real;
(4) an understanding of delirium (and hence madness) as a rationally structured *language*;
(5) a consequent realization that madness is a state of reason, but of reason in a condition of blindness;
(6) an analysis of this blindness as due not to a lack but an excess of light—that is, *dazzlement*;

(7)   a relation of this understanding of madness to the *fundamental Classical cosmology* of day and night (Gutting 1989, pp. 84–85).

The elements of this analysis are quite different from the elements that make up Foucault's analysis of the archaeology of the human sciences in *The Order of Things*. There are no organizing principles here, and little concern for the role of language and semiotics. Foucault's analysis here resembles more a literary analysis of the Classical conception of madness, than the archaeological analysis of *The Order of Things*. This is an archaeology of *experience*, of how people in a certain period viewed madness, how they thought about it and how they reacted to it. The approach is closer to phenomenology than Foucault's archaeology of knowledge. Foucault later regretted that experience was at the focal point in the book (Foucault 1972, p. 32).

In many ways *Madness* resembles more the later knowledge/power studies than the archaeology of *The Order of Things*: It is a history of practices and institutions, and of how the mad became excluded, subjugated and silenced. Only the concept of power is still the top-down power of repression, not the horizontal power of a struggle. Foucault developed his concept of power only when he was giving up on archaeology in the early 1970s, as we shall see in Chap. 3.

Like *Madness and Civilization*, *The Birth of the Clinic* is an archaeology of experience, or rather an archaeology of a *perception*—medical perception. But now there is an emphasis on the *discourse* of medical perception, i.e., there is a new emphasis on language, semiotics, and, most importantly, the historically contingent conditions of medical knowledge. The book opens with two observations. The first, from 1769, describes the effects of a cure of hysteria that consisted of "baths, ten or twelve hours a day, for ten whole months" (Foucault 1973, p. ix). In this description we read about "membranous tissues" from various organs peeling away like "damp parchment." The terminology is almost incomprehensible today; reading the description one has almost no idea what was actually being observed and one suspects that the description stems more from the doctor's imagination than from what he actually saw.

The second observation is from 1825 and it describes "an anatomical lesion of the brain and its enveloping membranes, the so-called 'false membranes' frequently found on patients suffering from 'chronic meningitis'" (Foucault 1973, p. ix). This description seems "modern": it could be from a recent medical textbook. What Foucault wants to know is how medical discourse changed; how it happened that doctors saw the human body in a completely different way within the space of a few decades. Foucault does not believe that what doctors used to observe was nothing but fantasy while later doctors chose to see what really was there. Foucault rejects the idea that at some point medicine replaced fantasy with objectivity and rationality. Rather, the 19th century discourse of medical knowledge was organized in a radically different way from that of the 18th century. Thus the most fundamental change happened at the end of the 18th century, at the break between the Classical age and the Modern age, when the individual human being became an object of scientific study:

> The *object* of discourse may equally well be a *subject*, without the figures of objectivity being in any way altered. It is this *formal* reorganization, *in depth*, rather than the abandonment of theories and old systems, that made *clinical experience* possible; it lifted the old Aristotelian prohibition: one could at last hold a scientifically structured discourse about an individual. (Foucault 1973, p. xiv)

This is a reorganization at the archaeological level, although Foucault still does not use the terminology of *The Order of Things*, let alone that of *The Archaeology of Knowledge*. The distinction between the surface knowledge ("theories and old systems") and the deep knowledge ("*formal* reorganization, *in depth*") plays a more prominent role here than in *Madness and Civilization*. Still, though Foucault talks about a mutation in the medical discourse, he neither identifies the deeper level where this mutation is to have taken place nor the nature of the rules or principles that change. Furthermore, in *The Order of Things* the metamorphosis of the subject into an object of scientific study is an effect of a mutation in the *episteme*, rather than an important feature of the mutation—it is the "origin" of man, the figure whose imminent demise Foucault predicts at the end of *The Order of Things*. In *The Birth of the Clinic* Foucault is clearly moving towards the archaeology of *The Order of Things*, but still retains the concern with institutions, practices and power. This concern is absent in *The Order of Things* and *The Archaeology of Knowledge*, but reappears in *Discipline and Punish*.

In *The Archaeology of Knowledge*, Foucault attempted to systematically lay out the archaeological method that had developed from *Madness and Civilization* through *Birth of the Clinic* to becoming an acknowledged method or type of study in *The Order of Things*. One of the most important features of archaeology is now made explicit, namely the autonomy of discourses. The history of discourses is not a history of who thought or wrote what, but a history of the discourses themselves—it is about the words and not their authors. The explanation for a mutation of discourse is to be sought in the underlying archaeological system, but neither in subjective influences of specific individuals nor in the *Zeitgeist*. For archaeology, man has already vanished. As Gary Gutting has pointed out, an archaeological investigation is an analysis of a system's structure that, *like* structuralist undertakings, does not seek explanation in the subject (Gutting 1989, p. 228). But *unlike* structuralist undertakings, archaeology does not put the structure itself and its possibilities in the center. Foucault is not interested in the formal possibilities of archaeological structures, but rather in their actualities and what makes the actualities possible:

> In contrast to those whom one calls structuralists, I am not so interested in the formal possi-bilities offered by a system like language. Personally I am rather haunted by the existence of discourse, by the fact that particular words have been spoken; these events have functioned in relation to their original situation, they have left traces behind them. (Foucault 1989a [1967], p. 25)

Until the publication of his *Archaeology of Knowledge* in 1970, Foucault was widely considered a structuralist, and may even have considered himself to be one at some point. He came, however, to resent being labeled a structuralist and denied vehe-mently being one. In the "Foreword to the English edition" of *The Order of Things*,

Foucault writes: "In France, certain half-witted 'commentators' persist in calling me a 'structuralist.' I have been unable to get into their tiny minds that I have used none of the methods, concepts or key terms that characterize structural analysis" (Foucault 1971, p. xiv). Of course Foucault did use some of the methods and many of the concepts and key terms of structural analysis, he even revised his *Birth of the Clinic* when it was to be reissued in 1972, and edited out much of the structuralist language (see Eribon 1991, p. 185, also pp. 167–168). It is true that the label 'structuralist' was given to very diverse work; in a 1969 interview Foucault reduces its use to a joke: "You know the joke: what's the difference between Bernard Shaw and Charlie Chaplin? There is no difference, since they both have a beard, with the exception of Chaplin of course" (Foucault 1989c [1969], p. 60).

Along with the systematization of archaeology comes a new terminology, but, after presenting it in *The Archaeology of Knowledge*, Foucault never used it again. Let me nonetheless give a brief sketch.

Foucault questions "traditional unities," such as disciplines, books, and works (*œuvre*). He wants to try to form new unities, not the *epistemes* or "deep" archaeological levels of *The Order of Things*, but "discursive formations":

> Whenever one can describe, between a number of statements, such a system of dispersion, whenever, between objects, types of statement, concepts, or thematic choices, one can define a regularity (an order, correlations, positions and functionings, transformations), we will say, for the sake of convenience, that we are dealing with a *discursive formation*. (Foucault 1972, p. 38)

Discursive formations thus consist of four kinds of elements: objects, statements, concepts and strategies; the elements are distributed, or dispersed, in a certain way and the discursive formation is defined by the regularity of this dispersion of the elements. There are rules that determine the formation of each of the elements, and these rules are not external to the discourse, nor a deeper level of it. Foucault places them variously "at the limits" of discourse or as "residing in discourse itself" (Foucault 1972, p. 74). The rules are groups of relations that determine the dispersion of elements on the surface of discourse. Discourse is now only surface, and archaeology is "*the project of a pure description of discursive events*" (Foucault 1972, p. 27).

Each kind of element has its conditions of existence and ordering within discourse alone. *Objects*, such as pathologies, types of individuals, classes of behavior etc. (Foucault is speaking of objects of knowledge, in particular objects of the human sciences), are formed through discourse: discourses "are practices that systematically form the objects of which they speak" (Foucault 1972, p. 49). The objects are not explained with regard to a pre-discursive reality, nor with regard to semantics; it is "neither by 'words' nor by 'things' that the regulation of the objects proper to a discursive formation should be defined" (Foucault 1972, p. 55).

As we can see from the last two quotes, Foucault sometimes speaks of objects ordered or regulated by discourse, and sometimes as formed, defined or constituted by discourse. The second formulation implies a rather radical nominalism, but Foucault does not seem to want to go quite that far. At one point in the *Archaeology of Knowledge* Foucault reminds the reader that: "we must not forget that a rule of formation is neither the determination of an object, nor the characterization of a type of enunciation, nor the form or content of a concept, but the principle of their multiplicity and dispersion" (Foucault 1972, p. 173; see also Foucault 1988 [1984], p. 257).

Statements is here short for "types of statement" or what Foucault also calls "enunciative modalities" or "enunciative types." Under this term he puts questions about the authority of the speaker (who is allowed to speak in the name of that discourse, as a scientist, as a doctor etc.), the institutional site of the speaker, and the relative position of the speaker within the discourse (e.g., as a speaking subject, listening subject, observing subject etc.). The distribution of places where statements can come from (their authority, institutional site and position within discourse) is a feature of the discourse: "it is neither by recourse to a transcendental subject nor by recourse to a psychological subjectivity that the regulation of its enunciations should be defined" (Foucault 1972, p. 55). What matters is not who the speaker is, but his authority, institutional site and discursive position.

As concerns *concepts*, the archaeologist describes "the organization of the field of statements where they [the concepts] appeared and circulated" (Foucault 1972, p. 56). Concepts are not to be reconstructed in a conceptual edifice, nor defined in terms of rational progress: "to analyze the formation of concepts, one must relate them neither to the horizon of *ideality*, nor to the empirical progress of *ideas*" (Foucault 1972, p. 63). Foucault describes the formation of concepts in some detail, but he is not only concerned with the formation of *concepts* and their conditions of existence within discourse, but also with *statements*: How they are grouped together, what makes them necessary, well-founded, assumed; the role of statements from other discourses (models, analogies, higher authority, necessary or generally accepted as true, etc.) and what determines their objectivity and even rationality. The implications of these rules are left largely unanalyzed. This is striking when one considers what they encompass, in light of the examples Foucault gives us. One example is that of the four schemata that characterize the general grammar of the Classical age, which he discussed in *The Order of Things*. The schemata are not a reconstruction of the conceptual system of general grammar; they make it possible to describe how:

> General Grammar defines a domain of *validity* for itself (according to what criteria one may discuss the truth or falsehood of a proposition); how it constitutes a domain of *normativity* for itself (according to what criteria one may exclude certain statements as being irrelevant to the discourse, or as inessential and marginal, or as non-scientific); how it constitutes a domain of *actuality* for itself (comprising acquired solutions, defining present problems, situating concepts and affirmations that have fallen into disuse). (Foucault 1972, p. 61)

The work that the schemata (or in general the rules of the formation of concepts) are supposed to do opens up vast fields within philosophy of science—just the question of validity points to a large volume of literature within Anglo-American philosophy of science on justification, standards of objectivity, styles of reasoning etc. Foucault's *domain of actuality* appears to be very similar to Kuhn's *paradigms*, but we get no further details of it. And the quote above describes only one of four kinds of tasks that the schemata perform. At least in the case of "concepts," Foucault draws together too many complex features of scientific knowledge that are to be described by rules or schemata that are too abstract and too vague to be useful.

By *strategies* Foucault means choices of themes and theories. In the 18th century, general grammar had the theme of an original language and the Physiocrats had a theory "of a circulation of wealth on the basis of agricultural production," (Foucault 1972, p. 64) to take just two examples from *The Order of Things*. Looking back on his three historical studies from the viewpoint of *The Archaeology of Knowledge*, Foucault notes that *Madness and Civilization* was concerned with the formation of objects (in particular madness as an object), *The Birth of the Clinic* was concerned with the formation of statements or "enunciative modalities" (the authority and position of the doctor; the development of medical institutions and so on), and *The Order of Things* was concerned with the formation of concepts (Foucault 1972, p. 65). Foucault still had to write a historical work where theory choice (or "strategies") were at the center, and therefore he claims to have little to say about them. He does, however, say that theory choice is not to be explained in terms of "fundamental projects," i.e., a foundational, true and timeless discourse on which other discourses rest; nor is it to be explained in terms of world-view, "play of opinions" or the external (e.g. social or economic) interests of the individuals involved. The formation of theoretical choices is to be explained in terms of the rules and regularities of the discourse under investigation and its relations to other discourses. The important point about the formation of theoretical and thematic choices is that one must look for the conditions that make *rival* theories possible within the same discourse. The archaeological work here is to determine the conditions that make a range of theoretical options possible.

Objects, statements, concepts and strategies are the elements of discursive formations, the unities Foucault wanted to replace the traditional unities of scientific disciplines, books and works. I have barely hinted at what the relations within and between these elements are supposed to tell us about discourses, but I hope to have conveyed at least some of the complexity and elusiveness of this theoretical construction. After studying it and the examples of its application, one is left with the impression that it does draw out in a systematic way structural similarities in Foucault's historical studies, but that it is left entirely unclear how it could apply to new examples and new studies. The explanatory value of this construction is also questionable: the object of analysis seems to be in a constant danger of disappearing in an endless web of relations, levels, rules, forms and functions. Archaeology seems to endlessly divide and fracture discourse, as Foucault himself realized: "Archaeological comparison does not have a unifying, but a diversifying effect." Its object seems to be in a constant danger of disappearing in details.

The innovative terminology and abstract theorizing of *The Archaeology of Knowledge*, together with other aspects of Foucault's work, found its way into the Anglo-American discourse of cultural theory and related compartments of philosophy, but has otherwise found little favour within philosophy let alone ethics. There have been some vigorous attempts to defend it in recent years (Tiisala 2015; Webb 2013), but I, for one, would wish for more archaeological studies, less theory. In my view, archaeology, as a method of analysis, stands and falls with its concrete applications.

References

Belon, Pierre. 1555. *Histoire de la nature des oiseaux*. Paris.

Dews, Peter. 1992. Foucault and the French Tradition of Historical Epistemology. *History of European Ideas* 14: 347–363.

Eribon, Didier. 1991. *Michel Foucault*. Trans. B. Wing. Cambridge, MA: Harvard University Press.

Foucault, Michel. 1961. *Folie et déraison: Histoire de la folie à l'âge classique*. Paris: Plon.

Foucault, Michel. 1965. *Madness and Civilization: A History of Insanity in the Age of Reason*. Trans. R. Howard. New York: Random House. This is an abridged translation of Michel Foucault's *Folie et déraison* (Paris: Plon, 1961).

Foucault, Michel. 1971. *The Order of Things: An Archaeology of the Human Sciences*. Trans. Alan Sheridan. New York: Random House. Originally *Les mots et les choses: une archéologie des sciences humaines* (Paris: Gallimard, 1966).

Foucault, Michel. 1972. *The Archaeology of Knowledge, and the Discourse on Language*. Trans. A.S. Smith. New York: Pantheon. Originally *L'archéologie du savoir* (Paris: Gallimard, 1969).

Foucault, Michel. 1973. *The Birth of the Clinic*. Trans. A.M. Sheridan Smith. New York: Random House. Originally *Naissance de la clinique: une archéologie du regard médical* (Paris: PUF, 1963).

Foucault, Michel. 1988 [1984]. The Concern for Truth. In *Politics, Philosophy, Culture: Interviews and other Writings 1977–1984*, ed. Lawrence D. Kritzman, 255–267. Trans. A. Sheridan. New York and London: Routledge. Originally published in *Le Magazine Littéraire* in 1984.

Foucault, Michel. 1989a [1967]. The Discourse of History. In *Foucault Live: Interviews, 1966–84*, ed. Sylvère Lotringer, 11–33. New York: Semiotext(e). This interview was originally published in *Les Lettres françaises* in 1967.

Foucault, Michel. 1989b [1969]. The Archaeology of Knowledge. In *Foucault Live: Interviews, 1966–84*, ed. Sylvère Lotringer, 45–56. New York: Semiotext(e). This interview was originally published in *Le Magazine Littéraire* in 1969.

Foucault, Michel. 1989c [1969]. The Birth of a World. In *Foucault Live: Interviews, 1966–84*, ed. Sylvère Lotringer, 57–61. New York: Semiotext(e). This interview was originally published in *Le Monde* in 1969.

Foucault, Michel. 2000 [1966]. The Order of Things. In *Essential Works Volume 2: Aesthetics, Method, and Epistemology*, ed. James Faubion and Paul Rabinow, 261–67. London Penguin Books. This interview was originally published in *Les Lettres françaises* in 1966.

Foucault, Michel. 2006. *History of Madness*. New York: Routledge. This is an unabridged translation of Michel Foucault's *Histoire de la folie à l'âge classique* (a revised edition of *Folie et déraison*) (Paris: Gallimard, 1972).

Gutting, Gary. 1989. *Michel Foucault's Archaeology of Scientific Reason*. Cambridge: Cambridge University Press.

Hacking, Ian. 1975. *The Emergence of Probability*. Cambridge: Cambridge University Press.

Hacking, Ian. 1979. Michel Foucault's Immature Science. *Nous* 13: 39–51.

Hacking, Ian. 1986 [1981]. The Archaeology of Foucault. In *Foucault: A Critical Reader*, ed. David Couzens Hoy, 27–40. Oxford and Cambridge, MA: Basil Blackwell. Reprinted from *The New York Review of Books*, 28:8 (1981): 32–37.

Koopman, Colin. 2010. Historical Critique or Transcendental Critique in Foucault: Two Kantian Lineages. *Foucault Studies* 8: 100–121.

Koopman, Colin. 2013. *Genealogy as Critique: Foucault and the Problems of Modernity*. Bloomington, IN: Indiana University Press.

McQuillan, Colin. 2010. Philosophical Archaeology in Kant, Foucault, and Agamben. *Parrhesia* 10: 39–49.

Ross, Alison. 2018. The Errors of History: Knowledge and Epistemology in Bachelard, Canguilhem and Foucault. *Angelaki* 23 (2): 139–154.

Sheridan, Alan. 1980. *Michel Foucault: The Will to Truth*. London and New York: Tavistock Publications.

Tiisala, Tuomo. 2015. Keeping It Implicit: A Defense of Foucault's Archaeology of Knowledge. *Journal of the American Philosophical Association* 1: 653–673.

Tiles, Mary. 1984. *Bachelard: Science and Objectivity*. Cambridge: Cambridge University Press.

Webb, David. 2013. *Foucault's Archaeology: Science and Transformation*. Edinburgh: Edinburgh University Press.

Chapter 3
Power, Knowledge, and the Politics of Truth

In my two case studies that follow this chapter, I rely heavily on Foucault's analysis of power/knowledge relations as I attempt to practice the sort of critique which I am here calling politics of truth. This chapter will therefore be focused on Foucault's idea of power, the relation between power and knowledge, the application of power/knowledge analysis to natural sciences, and the "politics of truth." In the first section I consider the development of the concept of power in Foucault's work of the early seventies and then how Foucault used it in *Discipline and Punish* to analyse the practice of incarceration and in *The History of Sexuality* to analyse attitudes toward sexuality. The second section is concerned with the relation between knowledge and power. I argue that although power/knowledge relations are diverse and must be studied in their particular and concrete details, there are two general (overlapping) types of power/knowledge relations that can be identified in Foucault's historical analyses. One has to do with how humans are classified, labelled and studied; the other has to do with *biopower*, i.e., how, on one hand, populations are managed ("biopolitics of the population"), and, on the other hand, how individuals are disciplined ("anatomo-politics of the human body").

In the third section I will argue that although Foucault's studies are concerned with those human sciences that might seem to be inexact, and even "immature" or "dubious," there is nothing that a priori excludes the natural (exact and mature) sciences from similar considerations; and, furthermore, there are good reasons to extend such considerations to the exact sciences, in particular biological sciences when they have humans as their subject. In the fourth and last section I discuss Foucault's use of the term "politics of truth" and conclude with a discussion of the critique for which I am using this term. My discussion of politics of truth will serve as a bridge between my discussion of Foucault and the two cases with which I shall be concerned in the two remaining chapters.

© The Author(s) 2018
G. Árnason, *Foucault and the Human Subject of Science*, SpringerBriefs in Ethics,
https://doi.org/10.1007/978-3-030-02813-8_3

3.1   Power

In *The Archaeology of Knowledge*, Foucault presented his archaeological method, complete with its own terminology and domain of research. It was an awkward construction and Foucault was disappointed. After the publication of *The Archaeology of Knowledge*, Foucault abandoned archaeological studies and left the new terminology for others to use and abuse. In the subsequent works there are no *epistemes*, no archaeological levels, no discursive formations, no enunciative functions, no empiricities, no positivities. There was a turning point: "The events of '68." It was of course nothing remotely close to another French revolution, but the student riots in Paris in May 1968 were a watershed for French politics and culture. French students took to the streets, built barricades and fought the police. Foucault was not in Paris in May 1968. He was in Tunisia, trying to fit together the theoretical pieces of his archaeology and in between reading Nietzsche on the beach. *The Archaeology of Knowledge* was published in 1969 and a year later, in September 1970, Michel Foucault became a professor at the Collège de France. He chose to name his chair "History of Systems of Thought," which points back to his archaeological work. But his inauguration address, published in French under the title *L'ordre du discours* and in English translation as *The Discourse on Language*, points also in a different direction. The field of his future work at the Collège de France was to be fixed by the following hypothesis:

> In every society the production of discourse is at once controlled, selected, organised and redistributed according to a certain number of procedures, whose role is to avert its powers and its dangers, to cope with chance events, to evade its ponderous, awesome materiality. (Foucault 1972, p. 216)

The result of this research plan was two books, *Discipline and Punish* and *History of Sexuality* (published in 1975 and 1976, respectively). I will here be concerned with the concept of power as it appears in these works, and in two lectures from 1976 which in English translation are simply titled "Two Lectures" (Foucault 1980a [1976]). The concept of power that emerged in Foucault's work in the early- and mid-seventies is perhaps most clearly formulated in "Two Lectures," but it really comes to life only through its application to problems such as incarceration and sexuality.

Foucault's "power" is, in keywords, *horizontal, diffuse, dynamic, active* and *creative*. Power is *horizontal*: it is not the top-down power of the sovereign, but is *diffused* through nearly all human relations. The metaphor of network is more appropriate than the metaphor of hierarchy. Power is *dynamic*: it is not a right or privilege someone possesses; it exists only when it is exercised. No one possesses power, but everyone exercises it. Power is *active*: there are no passive roles in power relations, since power is manifested in struggle rather than repression. And, finally, though it may be repressive, power is more importantly *creative*: it is like electricity that turns the wheels of society and culture, and therefore of science as well. This description of Foucault's notion of power is cursory and metaphorical, but it is meant only as an introduction. All these aspects of Foucault's power will be discussed further and clarified in the course of my discussion below of the application of the concept in Foucault's power/knowledge studies.

Foucault's idea of power does not exclude or replace the common idea of top-down political power. On Foucault's picture of modern Western societies there is still political power with its centres in governments and state institutions. There is of course also the brute power of authorities as well as individuals. Foucault's power is meant to capture another kind of power; subtle, diffuse, pervasive, horizontal power, which has its historical beginnings in 17th-century Europe.

Foucault labels his notion of power *Nietzschean* in his "Two Lectures" and contrasts it with what can be recognized as the Marxist notion of power, which he labels *Reichean* (Foucault 1980a [1976], p. 91). The second notion is that of 19th century economics, which Foucault insisted was originated by Reich and merely elaborated by Marx. Foucault's notion is, however, only partly Nietzschean. Essential to Nietzsche's notion of power are the manifold forces of life; for him power is just the force internal to living beings to grow and get stronger. Any consideration of power could not be separated from considerations of health, sickness, growth, decadence, climate, diet, etc. Nietzsche's notion of power is *physiological*, in the sense that it has always to do with the nature, state and health of a living being. Foucault's power is *relational*, in the sense that it has always to do with one's relations to others or oneself, it has the effect of constraining thought, action and even being (i.e., affecting what one can do, what one can think and what one can be). Yet, what the two notions have in common is that they take power out of the strictly political context of sovereign or juridical power and give it a more general application.

In the early seventies Foucault's notion of power was already taking form as being dispersed and pervasive, and being based not on direct physical power but subtle controls through, for instance, surveillance—in 1972 Foucault was already studying Jeremy Bentham's plans for the Panopticon (Foucault and Deleuze 1977, p. 210). At this time, however, Foucault considered power primarily repressive and, by definition it seems, a bad thing needing to be fought. Despite his disagreements with communism, Foucault's discussion of power seems to have taken place more or less in the context of revolutionary Marxism, as is borne out by interviews and discussions published in that period [e.g. "Rituals of Exclusion" (Foucault 1989a [1971]); "Revolutionary Action: 'Until Now'" (Foucault 1977a [1971]); "Intellectuals and Power: A conversation between Michel Foucault and Gilles Deleuze" (Foucault and Deleuze 1977 [1972]); and "On Popular Justice: A Discussion with Maoists" (Foucault 1980b [1972])].

Foucault accepted that the proletariat is in a struggle against bourgeois oppression, but adds that other groups are also struggling against their oppressors: "Women, prisoners, conscripted soldiers, hospital patients, and homosexuals have now begun a specific struggle against the particularized power, the constraints and controls, that are exerted over them." Those groups will join the proletariat in the revolution "because power is exercised the way it is in order to maintain capitalist exploitation" (Foucault 1977a, p. 216).

In the interviews and discussions referred to above Foucault takes the logical step from the view that power is pervasive, repressive and bad, to anarchism—without ever using the word. After the revolution, or whenever our fight against local and global powers has been won, the old power structures cannot be simply replaced with new ones serving the same function; the old judicial system, police and army cannot just be replaced with another judicial system, police and army, for such systems are always formed by and for the bourgeoisie. A society without power, without control, without authority, can only, it seems, be an anarchist non-society. In one discussion with young would-be revolutionaries, Foucault, taking issue with their use of the phrase "the whole of society," remarks: "'The whole of society' is precisely that which should not be considered except as something to be destroyed. And then, we can only hope that it will never exist again" (Foucault 1977a [1971], p. 233). And in a discussion with Maoists, following one Maoist's claim that during the first stage of the revolution looting and killing is just fine because "the stick must be bent in the other direction," Foucault adds that "above all it is essential that the stick be broken" (Foucault 1980b [1972], p. 32).

In a discussion with Noam Chomsky on the Dutch television network NOS in 1971, Foucault managed to shock Chomsky by arguing that the very notions of justice and human nature might have to be abandoned along with the capitalist society as having been "invented and put to work in different types of societies as an instrument of a certain political and economic power or as a weapon against that power" (Achbar 1994, pp. 31–33). Chomsky later claimed that Foucault was the most amoral person he ever met (Miller 1994, p. 201).

As his thinking about power developed, Foucault retreated from his apparent anarchist position but his aversion to Marxism remained. In his *Discipline and Punish* Foucault gives a nod to Marxist analysis and then gives it his own twist: "it is largely as a force of production that the body is invested with relations of power and domination; but, on the other hand, its constitution as labour power is possible only if it is caught up in a system of subjection" (Foucault 1977b, p. 26). It is the system of subjection that matters, not labour, production or surplus value.

In *Discipline and Punish* Foucault pictures society as a highly organized system of power relations, where the penal system serves as one technology of power among others. Foucault's study of the history of penal practices is not a history of punishment from the perspective of law or with respect to the evolution of legislation, nor is it merely a history or study of the social forms (as in Durkheim's studies); it is not a study of the justification or legitimacy of punishment, yet it combines aspects of and has a bearing on all of these fields of study.

Foucault liked to illustrate changes in scientific thought and social practices by contrasting texts from two periods not very far from each other, manifesting drastically different way of thinking or doing things. The two texts at the beginning of *Discipline and Punish* serve just this purpose: First, there is a description from 1757 of a public execution in Paris and then there is a time table from 1838 regulating in detail the daily activities of young prisoners in Paris. It is an invitation to think about why and how a society moves over a fairly short time from practicing punishment as a gruesome public spectacle, to punishing people by locking them up and subjecting them to carefully structured routines. Similar juxtapositions of texts in Foucault's archaeological work seemed at first glance to show that scientists, scholars or medical doctors had in earlier times been irrational and prone to fantasy and erroneous observations, but over a relatively short period of time became accurate in their observations and rational in their production of knowledge. Foucault's archaeological stories are about transformations of "systems of thought," which involve a change in standards of rationality and not a movement from irrationality to rationality. The transformation of penal practices from the 18th to the 19th century may seem at first glance to be a change from barbaric cruelty to more humane practice of punishment. On Foucault's account, this transformation was not the result of reforms intended to make punishment more humane, the reforms were intended to make punishment more effective and useful for the management of the population. The main objectives of the judicial reforms were

> to make of the punishment and repression of illegalities a regular function, coextensive with society; not to punish less, but to punish better; to punish with an attenuated severity perhaps, but in order to punish with more universality and necessity; to insert the power to punish more deeply into the social body. (Foucault 1977b, p. 82)

This does not mean that there were any overarching plans to transform society. Foucault's account is one of incidental alignment of various strategies and plans, of local needs and actions, that come together in a certain "economic rationality," that just happens to have the appearance of humane reforms (Foucault 1977b, p. 92).

Foucault's pessimistic, even cynical, view of the penal reforms puts penal practices in the context of social control, as opposed to the context of justice and rehabilitation. It furthermore makes the disillusioning genealogical point that the origin of the modern penal system is not "pure" or a necessary conclusion of the rationality of human progress, but contingent and haphazard. Foucault thus rejects any teleological account of the emergence of modern penal practices.

Foucault is writing a "history of the present," in the sense that he offers a story about how the present power structures historically developed out of simple and less effective organizations of power as well as an analysis of how power is now organized. Here the former story informs the latter: we understand better how power is presently organized through a story about how it came to be organized the way it is. Foucault's analysis of power and stories about its development do not make it clear what it is that fuels the transformation from a simple, relatively ineffective power structure to a complex, efficient network of power. Power networks appear to emerge and grow by virtue of some kind of an internal force, as if it is in the nature of power networks to grow. It is as if there is a Nietzschean "will to power" built into the power networks: they tend to develop towards higher complexity and more efficiency just because that is how they grow, get stronger, increase their power. Foucault offers little in way of explanations, but the picture that appears in the description of power networks is that power networks grow through incidental and contingent convergence of a multitude of plans, strategies and pressures at the micro-level of power relations.

The first volume of *The History of Sexuality* is also concerned with a transformation, in this case how sexuality became what it is today. Foucault sets out to reject what he calls "the repressive hypothesis." The repressive hypothesis is a familiar story of how sexuality was repressed in Europe in Victorian times, and how it was then liberated during the 20th century. Foucault explains how the repression of sexuality in the 19th century did not consist in ignoring sexuality or putting it somehow outside of our thought and discourse. On the contrary, the repression of sexuality was an aspect of an intense concern with it. The result was not that sexuality was invisible or ignored, but that sexuality was everywhere. Foucault's gives us examples, including changes to sexual confession in the Catholic church, which gained importance, "everything had to be told" (Foucault 1978, p. 19). Schools where another institutional locus for concerns about sexuality:

> The space for classes, the shape of the tables, the planning of the recreation lessons, the distribution of the dormitories (with or without partitions, with or without curtains), the rules for monitoring bedtime and sleep periods—all this referred, in the most prolix manner, to the sexuality of children. (Foucault 1978, p. 28)

Foucault claims that there is no biological function of sex which is the basis of sexuality. On the contrary, sexuality gave rise to the notion of sex "as a speculative element necessary to its operation" (Foucault 1978, p. 157). Unfortunately Foucault does not go into any detail of how sex is historically constituted by the deployment of sexuality.

In the first volume of the *History of Sexuality*, Foucault has very little to say about sexuality before the end of the 17th century. Foucault does tell a story of a transformation of sexuality, but he does not have much to say about the old power structure out of which the new one emerged except for the occasional references to Catholicism and the practice of confession. This silence was not because he thought there was not much to say about it, but because he thought there was so much to say about it that he was going to save it for later volumes of the series. The book on sexuality in the Renaissance period and up to mid 17th century, *The Body and the Flesh*, was written but not published until 2018. A book was published on sexuality in Ancient Greece, *The Use of Pleasure* and another on sexuality in Hellenistic and Roman times, *The Care of the Self*. Both were published posthumously and there Foucault had shifted the focus from power to the self, that is, the individual and his concern with himself.

Foucault's *History of Sexuality* is an analysis of the emergence of a power network which transforms sexuality to make it useful. Foucault goes into considerable detail of how everything was sexualized, so to speak, but it is not clear in what sense this transformation was useful nor is it clear what drove it. The story is convincing, in so far as Foucault shows how sexuality was not repressed but magnified, multiplied and diversified; and one sees power exercised in the endless details of the discourse on sexuality: in talk, in silence, in rules and prohibitions, in organization of institutions, in public campaigns, in new sexual pathologies and treatments, and so on and so forth. The story is less convincing when it comes to the usefulness of the transformation of sexuality, but it seems to have to do with the rise of the efficient power mechanisms in Western societies during the 18th and 19th centuries. Through industrialization there was a growing need to manage populations. The workforce had to be managed and organized in order to maximize its use for production:

> Through the political economy of population there was formed a whole grid of observations regarding sex [...] It was essential that the state know what was happening with its citizens' sex, and the use they made of it, but also that each individual be capable of controlling the use he made of it. Between the state and the individual, sex became an issue, and a public issue no less; a whole web of discourses, special knowledges, analyses, and injunctions settled upon it. (Foucault 1978, p. 26)

As the governance of the social body becomes more efficient, sexuality becomes an issue, certainly in order to manage birth-rates but also as something that needed to be managed for its own sake:

> This was the first time that society had affirmed, in a constant way, that its future and its fortune were tied not only to the number and the uprightness of its citizens, to their marriage rules and family organization, but to the manner in which each individual made use of his sex. (Foucault 1978, p. 26)

This management of sexuality is an important example of biopower, both with regard to the management of populations and with regard to the disciplining of the body (the notion of biopower will be further discussed in the next section). According to Foucault (1980c [1977], p. 125), sexuality is where the discipline of the body and the management of populations intersect.

It is worth emphasizing that what emerges out of Foucault's study of power relations is not a *theory* of power. Foucault has an original *concept* of power, which he applies in his power-analyses, but he has no universal theory of power. What emerges out of Foucault's analyses is not a theory, but an analysis of the power relations specific to the particular field of study, be it penal practices or sexuality (Foucault 2007 [1978], pp. 1–2; May 1993, p. 2; Koopman 2013, pp. 37–38). As a result, Foucault does not have a political theory, nor a global plan to resist power. Instead of politics he offers "micro-politics," instead of global action against perceived centres of power he analyses local struggles within specific power relations.

3.2 Knowledge

In the *Archaeology of Knowledge*, Foucault gave the word "institution" the wide sense of "that which is not said", as opposed to that which is said. That is, institutions are the non-discursive aspects of society, what is outside of discourse (Foucault 1980c [1977], pp. 197–198).

His archaeology was concerned with discourse, but the analysis of power and knowledge covered not only discourse but also institutions (in this wide sense of the social outside discourse). Foucault's archaeology was concerned with the principles that organize knowledge at different times and the knowledge-structures that they make possible. His concern was with (more or less) autonomous discourse, with the totality of texts and not particularly with the individuals who wrote the texts. The object of his studies was texts, utterances, what had been said (or rather, written)—in short, discourses. His two early archaeological books, however, also considered knowledge in relation to certain institutions, the asylum in *Madness and Civilization* and the clinic or hospital in *The Birth of the Clinic*, but the *magnum opus* of Foucault's archaeological period, *The Order of Things*, is not concerned with institutions. The early concern with institutions was recast in the context of power relations: Foucault studied the strategies of power that resulted in the modern prison and penal practices in *Discipline and Punish*, and in the first volume of *The History of Sexuality* he studied the strategies of power that transformed the way we deal with sexuality and how that transformation is manifested in the arrangements of a number of institutions, such as the school, the church and the family.

Foucault was already concerned with the relation of power and knowledge by the end of the sixties, for instance in his *Discourse on Language* and a little later in his wonderful essay "Nietzsche, Genealogy, History," (Foucault 1972, p. 216, 2000 [1971]) but his idea of this relation developed as his view of power itself developed during the early seventies. It was of course no news that "knowledge is power," but the relation between knowledge and power is much more complex than what Francis Bacon's phrase suggests. Obviously, and Foucault found it necessary to stress this, there is not a relation of identity between the two, knowledge is not the same as power.

> If I had said, or wanted to say, that knowledge was power I would have said it, and having said it, I would no longer have anything to say, since in identifying them I would have had no reason to try to show their different relationships. (Foucault 1989b [1984], p. 304)

Rather, power and knowledge always go together. There is no power without knowledge and no knowledge without power.

> There can be no possible exercise of power without a certain economy of discourses of truth [e.g. scientific knowledge] which operates through and on the basis of this association. We are subjected to the production of truth through power and we cannot exercise power except through the production of truth. (Foucault 1980a [1977], p. 93)

Knowledge and power are not the same, but, at least in our modern Western societies, one cannot exist without the other (Koopman 2013, p. 37). The relationship between power and knowledge is poorly described in abstract terms; there isn't even any single universal relation of power and knowledge. To understand power-knowledge relations, one must study particular sets of them in their concrete historical details. Still, there are at least two sorts of general relationship between power and knowledge, one concerned with classification, labelling and studying humans, the other concerned with "biopower." The two are surely not distinct, but I will discuss them separately before looking at their relation to each other.

As an example of a general power/knowledge relation of the first kind, let me consider the transformation of penal practices in late 18th century according to Foucault's analysis. A new system of power (in keywords: pervasive, subtle and efficient, instead of direct, brute and inefficient) required a whole new way of thinking about crime, punishment and criminals. Now it was necessary to know not only the criminal act but also the criminals and delinquents themselves in order to punish, correct and discipline them. The offence is fixed within an elaborate discourse of knowledge—that is, scientifically—and the offender is fixed within the same discourse as a

delinquent. The delinquent is not something that existed before the modern penitentiary apparatus came to existence, waiting to be discovered and dealt with. The two
emerged and developed together, since the delinquent himself was formed through
the techniques and methods of the penitentiary apparatus. Modern penitentiary power
creates the delinquent as its object and then strives to know this object through relent-
less examinations, interrogations, studies, biographies, reports and records: "it is this
delinquency that must be known, assessed, measured, diagnosed, treated when sentences are passed" (Foucault 1977b, p. 255).

Knowledge about the delinquent is important in order to deal with him properly,
but the "will to knowledge" built into the penitentiary apparatus does not merely
value knowledge as means to technical mastery of its object. The delinquent must be
understood in order to be judged at all. It became unthinkable to judge a person not
sufficiently known and understood (Foucault 2002b [1978]). And when the delinquent is known as an anomaly, as an abnormality, as a deviation, this knowledge
may call for "treatment," no less than the need for treatment may call for knowledge.
New knowledge creates new objects and new needs to deal with the objects—and
the study of the objects and dealing with them in turn creates new knowledge.

The emphasis here is on knowledge of people, as opposed to knowledge of nature.
It is often claimed, and often with reference to Francis Bacon, that the motivation,
even *raison d'être*, of science is the harnessing of nature through technology—we
are supposed to want to know nature in order to increase our power over it. But
Foucault's relation of power and knowledge is not as simple as that of knowledge
for technological mastery. His concern was with the emergence of modern power
formations as a way of having society efficiently govern itself as a mesh of horizontal power networks instead of being governed through the direct top-down power
of the king. These modern power formations happened to require certain kinds of
knowledge production to function, and the knowledge production in turn affects the
power formations. Foucault is here only concerned with the "sciences of man," but
it would certainly be worth investigating to what extent the natural sciences served
the modern power formations.

The rise and development of efficient power formations in early 19th century is
intertwined with new forms of knowledge and an increased emphasis on producing
knowledge about people. Foucault offers quite a few historical examples of this
transformation. One of the most memorable ones is the story of a poor, simple-
minded farm hand who gave little girls a few pennies for "caressing" him (or for
playing a game called "curdled milk") and, in 1867, was turned in to the authorities
by one girl's parents. The intolerance of such acts was probably not new, but the
treatment he got was new: The parents pointed him out to the mayor of the village,
he was then "reported by the mayor to the gendarmes, led by the gendarmes to the
judge, who indicted him and turned him over first to a doctor, then to two other
experts who not only wrote their report but also had it published" (Foucault 1978,
p. 31). In not so much earlier times the man would perhaps have been chased away
or suffered the anger of the parents, if anything had happened at all. At this point,
however, it had become necessary to study him. At this point the man and his actions
had become

the object not only of a collective intolerance but of a judicial action, a medical intervention, a careful clinical examination, and an entire theoretical elaboration. The thing to note is that they went so far as to measure the brainpan, study the facial bone structure, and inspect for possible signs of degenerescence [in] the anatomy of this personage who up to that moment had been an integral part of village life; that they made him talk; that they questioned him concerning his thoughts, inclinations, habits, sensations, and opinions. And then, acquitting him of any crime, they decided finally to make him into a pure object of medicine and knowledge—an object shut away till the end of his life in the hospital at Maréville, but also one to be made known to the world of learning through a detailed analysis [...] So it was that our society—and it was doubtless the first in history to take such measures—assembled around these timeless gestures, these barely furtive pleasures between simple-minded adults and alert children, a whole machinery for speechifying, analyzing, and investigating. (Foucault 1978, pp. 31–32)

This man not only needed to be dealt with and treated (although medically and not judicially as it turned out), he also became valuable as a "pure object of knowledge." It had become important and interesting to get knowledge about this "type" of a human being. A kind of person was emerging, what today is referred to as the sex-offender.

Further examples of this explosion of knowledge production can be found in two "dossiers" edited by Michel Foucault, one concerning punishment, the other sexuality. The first, published in France in 1973, concerns a man, Pierre Rivière, who murdered his mother, sister and brother in their home in Normandy on June 3rd, 1835. Parricides were not completely unheard of, either then or before, but the reaction Pierre Rivière got from society was quite different from what it would have been only a few of decades earlier. Pierre Rivière was sentenced to death and subsequently the sentence was commuted to life imprisonment (he hanged himself in Beaulieu prison in 1840). But Pierre Rivière was not only punished, he was studied: The "dossier" includes Rivière's memoir, medico-legal opinions, reports and records of examinations and interrogations [one fragment: "Q.: ... you have often crucified frogs and young birds; what is the feeling that led you to do such things? A.: I took pleasure in them" (Foucault 1975, p. 35)].

It was not enough to establish the crime and that Rivière committed it, he could not be sentenced before his character, desires, upbringing and nature had been understood. Foucault dates the beginning of the discourse on delinquents in the 1830s, which makes Rivière's case one of the first to be taken up into a discourse of truth about delinquents (Foucault 1989c [1976], p. 131).

The second dossier, published in France in 1978, concerns a hermaphrodite, Herculine Barbin, who was brought up as a female but reclassified as a male by a court judgment on June 21, 1860 (Foucault 1980d). The book contains (again) a memoir of the protagonist, as well as various documents including medical reports with detailed accounts of medical examinations before and after Barbin's death. One learns how Barbin's ambiguous sex poses a problem that schools, hospitals and courts must deal with. The discovery of this ambiguity and the need to deal with it required the determination of his/her "true sex," which in turn required considerable knowledge to be produced about him/her and his/her kind of case.

The explosion in demand, production and circulation of knowledge about people, to use metaphors from economics, took place at the same time as the emergence of what Foucault termed biopower. Knowledge about individuals and "human kinds" is one sort of a general power/knowledge relation, biopower is another.[1]

"Biopower" is one of those memorable terms that Foucault introduced and then quickly abandoned, but that still came to be seen by many as central to his work.[2] It appears in the first volume of the *History of Sexuality*, where it occurs only five times (Foucault 1978, pp. 138–145; see also Rabinow and Rose 2006, p. 199), and it is not used at all in the two subsequent volumes of the *History of Sexuality*. Foucault does not seem to have used the term in his lectures at the Collège de France, with two exceptions. On 17 March 1976, in his last lecture that year, Foucault briefly discusses biopower along the same lines as in the first volume of the *History of Sexuality* (Foucault 2003 [1976], pp. 243, 253–263). The other exception is his next lecture, on 11 January 1978, after a one year sabbatical, where he starts the lecture with the words: "This year I would like to begin studying something that I have called, somewhat vaguely, bio-power" (Foucault 2007 [1978], p. 1). Yet, he did not use that term again in his lectures and his concerns with the types of power he referred to as biopower became subsumed under the terms "government" and "governmentality" over the next three years (Foucault 2007 [1978], 2008 [1979], 2014 [1980]). The three-volume collection of essays, interviews and other minor texts, published as *Essential Works of Michel Foucault 1954–1984*, does not have a single index entry for the term biopower. Thus the term itself cannot be said to have much weight in Foucault's work, but it is a catchy term and it covers two areas or forms of power/knowledge relations, which Foucault discussed extensively throughout the second half of the 1970s.

In the first volume of the *History of Sexuality*, Foucault terms the two types of power, which together make up biopower, "anatomo-politics of the human body" and "bio-politics of the population" (Foucault 1978, p. 139). In his last lecture of 1976, and the first lecture of 1978, he uses biopower somewhat ambiguously to refer to either "bio-politics of the population" or the union of the two forms of power, where the first has been fully integrated into the second. It is useful to keep the term biopower for the integrated forms of power, and use biopolitics for the "bio-politics of the population," although Foucault himself was not entirely consistent in how he used those terms.

Of the two types of power, anatomo-politics of the human body developed first. Its beginnings are in the seventeenth century with industrialization but it only took full shape during the 18th century. Its focus was on the human body: "its disciplining, the optimization of its capabilities, the extortion of its forces, the parallel increase of its usefulness and its docility, its integration into systems of efficient and economic controls" (Foucault 1978, p. 139). In *Discipline and Punish* Foucault discusses the

---

[1] The idea of human kinds has been investigated by Ian Hacking, for instance in Ian Hacking's (1986), "Making up people." Hacking later proposed the term "interactive kinds" instead of "human kinds," see Hacking (1999, p. 103ff). Hacking has rejected that label as well, claiming that there is no such distinct and definable class of people that can be called a "human kind" or "interactive kind," see Hacking (2007).

[2] As has been pointed out (Esposito 2008, pp. 16–18), Foucault did not coin the term biopower, its use goes back at least to 1905.

body pole of power as *disciplinary power*. He offers a memorable example of the transformation of soldiers from being naturally soldier-like, strongly built and coura-geous, to ordinary men being made soldiers through training and discipline (Foucault 1977b, p. 135).

The bio-politics of the population developed somewhat later than the anatomo-politics of the human body, starting in the eighteenth century. It was concerned with humans not as bodies but as species; it was concerned with human populations, or in Foucault's words: "propagation, births and mortality, the level of health, life expectancy and longevity, with all the conditions that can cause these to vary" (Fou-cault 1978, p. 139, see also 2003 [1976], pp. 242–247). Various techniques emerged to intervene in and control populations, which required the production of enormous amounts of knowledge about populations. This knowledge production led not only to the birth of demographic statistics in the 18th century, but fuelled the development of statistics as an entirely new 'style' of scientific reasoning (Hacking 1992).

Foucault describes the two forms of power, the anatomo-politics of the human body and the bio-politics of the population, as "two poles of development linked together by a whole intermediary cluster of relations" (Foucault 1978, p. 139). The two poles were separate in the 18th century, but merged in "the great technology of power" in the 19th century. In the 18th century, we have on the one hand anatomo-politics of the human body manifested in "institutions such as the army and the schools, and in reflections on tactics, apprenticeship, education, and the nature of societies," and on the other hand bio-politics of the population manifested in "the emergence of demography, the evaluation of the relationship between resources and inhabitants, the constructing of tables analysing wealth and its circulation: the work of Quesnay, Moheau, and Süssmilch" (Foucault 1978, p. 140).

Biopower, as the disciplinary techniques of the human body and the regulation of populations, is therefore only relevant to knowledge/power analysis of the life sciences and biotechnologies to the extent that the life sciences and biotechnologies, and their products, have a role in disciplining the body or controlling populations.

The two general types of relationship between knowledge and power that I have described here are neither universal nor necessary, they are contingent features of the organization of power in modern societies. The analysis of these, and other, power/knowledge relations can be a form of critique, but not because these relations are necessarily or inherently bad. Historical accounts of the emergence or develop-ment of certain power/knowledge relations can help us understand a situation, in which we find ourselves. It is not a matter of unmasking ideologies and it is not a matter of liberation from power. Power is not always oppressive, harmful or evil. Understanding the relations of power and knowledge in a specific, concrete case can be an important step towards resistance, rather than towards liberation. In the next two sections, I hope to show how the analysis of power/knowledge relations can form a critique, particularly when we are concerned about human subjects of science.

3.3 Sciences of Man, Sciences of Nature

The scientific knowledge in my discussion so far has been the product of the "human sciences," simply because Foucault's studies were about them: In *Madness and Civilization* about psychiatry, in *The Birth of the Clinic* about medicine, in *The Order of Things* about three fields of study concerned with life, labour and language, in *Discipline and Punish* about the sciences of the penal system and in his *History of Sexuality* about the sciences of sexuality. Since the last two books are only about sciences in a wide sense, perhaps it would be more exact to talk about technologies and knowledges tied up with penal practices and sexuality.

It is not the case that Foucault avoided the natural sciences because he thought his methods did not apply to them. Rather, the human sciences suited him better as he considered them more enmeshed in society and politics than the exact sciences:

> If, concerning a science like theoretical physics or organic chemistry, one poses the problem of its relations with the political and economic structures of society, isn't one posing an excessively complicated question? Doesn't this set the threshold of possible explanations impossibly high? But on the other hand, if one takes a form of knowledge (*savoir*) like psychiatry, won't the question be much easier to resolve, since the epistemological profile of psychiatry is a low one and psychiatric practice is linked with a whole range of institutions, economic requirements and political issues of social regulation? Couldn't the interweaving of effects of power and knowledge be grasped with greater certainty in the case of a science as "dubious" as psychiatry? It was this same question which I wanted to pose concerning medicine in *The Birth of the Clinic*: medicine certainly has a much more solid scientific armature than psychiatry, but it too is profoundly enmeshed in social structures. (Foucault 1980c [1977], p. 109)

Furthermore, Foucault did not think the natural sciences particularly concerned him. He was happy with people from other disciplines using his methods and strategies, but he just did not think that he had offered any theories that applied equally to all domains of scientific knowledge (Foucault 1980e [1976], pp. 64–67). He did offer some "modes of approach" that might be useful for different disciplines or different domains of knowledge, but these modes did not constitute a theory, and definitely not a universal theory.

In an interview conducted in 1978, Foucault is asked whether his power/knowledge analysis concern the exact sciences. His initial reaction was to exclaim "Oh no, not at all! [...] I'm an empiricist: I don't try to advance things without seeing whether they are applicable" (Foucault 1988 [1984], p. 106). But then he adds that there does seem to be something general to be said about scientific knowledge, which applies then to the exact sciences, namely that (exact) science has been institutionalized as a power:

> It is not enough to say that science is a set of procedures by which propositions may be falsified, errors demonstrated, myths demystified, etc. Science also exercises power: it is, literally, a power that forces you to say certain things, if you are not to be disqualified not only as being wrong, but, more seriously than that, as being a charlatan. Science has become institutionalized as a power through a university system and through its own constricting apparatus of laboratories and experiments. (Foucault 1988 [1984], pp. 106–107)

Foucault characterizes the power of the natural sciences with reference to two sorts of institutions and one sort of power effects. The two sorts of institutions are the university on one hand, and laboratories and experiments on the other. An analysis of how science has become a power institutionalized in the university will encounter the question of the value of truth, or in Nietzschean terms, the question of the will to truth. We give truth great value; we usually prefer truth to lies, illusion or myth; often we associate, if not equate, truth with freedom and justice. Producing and circulating truths is imperative in modern societies. And nothing is taken to have a stronger claim to truth in modern societies than the products of science: scientific knowledge. The natural sciences have their roles and characteristics as a will to truth institutionalized in the university, and hence as a form of power.

The institutional constraints of laboratory and experiment, and scientific practice more generally, is something that Foucault never dealt with, as he was not concerned with the natural sciences. This has, however, turned out to be a very fertile field for students of science, philosophers as well as historians, sociologists, and anthropologists, especially in the last two decades of the 20th century. One of the main issues that have come up in studies of scientific practice is the emergence of scientific objects. Andrew Pickering, to take one example, wrote a book on the social construction of quarks (Pickering 1984). Not everyone who studies scientific practice is a social constructivist, but there is a strong tendency in the field to explain the emergence of scientific objects, even quarks, through the (often social) constraints of laboratory and experiments. I will not go into any detail of this issue or the controversies that ensued, I only want to make the point that studies of "the constraints of laboratory and experiment," which Foucault merely mentioned once in passing, have led many researchers to say similar things about the emergence of the objects of natural sciences as Foucault said about the emergence of the objects of the human sciences and I paraphrased above: new knowledge creates new objects and new needs to deal with the objects, which in turn gives rise to new knowledge. It does not follow that Foucault would have held that view if he had applied his power/knowledge analysis to natural sciences.

Those who would say things like "new knowledge creates new objects" might be called nominalists. But even if one rejects nominalism about the objects of the natural sciences, there are still epistemologically significant things to be said about the relation between power and knowledge of the natural sciences sort. I am not making a nominalist claim about the objects of the natural sciences. But I do claim that there is a relation between power and knowledge that goes not only from knowledge to power (that knowledge is used and has all sorts of effects) but also back from power to knowledge (that knowledge is demanded, produced and circulated)—and if the first direction of the relation leads us from epistemology to politics then the reverse direction surely leads us back to epistemology. In other words, it is not the case that fundamentally neutral and objective knowledge is used and abused for political ends, but rather that, because of how scientific knowledge and practice is interwoven in the very fabric of our society, epistemology and politics cannot be separated without one or both withering and dying.

In the quote above, Foucault alludes to only one kind of power effect that natural sciences can have, namely that of disqualifying, silencing and even ridiculing or excluding those who do not utter the right sort of sentences. This is surely not the only sort of power effect natural sciences can have, although it is an important one. The natural sciences can determine, or at least affect, not only what sort of things one can say and think, but also what one can do and what one can be. This will be my main concern here regarding the relation of natural sciences to power, in particular how natural science can affect what one can be, that is, what it means to be human and what different ways there may be to be human.

Joseph Rouse has given a particularly interesting account of the issue of power in the natural sciences. Drawing on Foucault's concept of power and Heidegger's idea of practice, he argues in his *Knowledge and Power* that the natural sciences are not exempt from the problematics of politics and hermeneutics, which some thinkers, such as Charles Taylor, Jürgen Habermas and Hubert L. Dreyfus, have considered the defining characteristics of the human sciences; and furthermore, the natural sciences differ from the human sciences primarily in that the former contain, for historically contingent reasons, specific techniques of power which the latter lack (Rouse 1987, see in particular Chaps. 6 and 7). These techniques of power have to do with laboratories and experiments, as Foucault already suggested in the quote above. As a consequence, the natural sciences should not be exempt from the sort of politically critical stand which is commonly taken with regard to the human sciences.

Let me take two examples of how natural sciences can affect our way of being, the first is from Joseph Rouse, the second from Hannah Landecker. One way in which natural sciences affect what we are, according to Rouse, is simply by forcing us to view ourselves as a part of nature, as essentially natural beings to which all the same rules apply as to other natural beings, as essentially just another kind of animal. This has at least two implications. First, there is a drive to naturalize our perspectives on the self and the society. We see ourselves as just one part of nature, and thus our understanding of nature shapes our self-understanding. Our understanding of nature stems from the natural sciences, which then must shape our self-understanding. Furthermore, we try to apply the same methods studying ourselves that we use to produce knowledge about nature; we try to explain human phenomena in the same way we try to explain natural phenomena. The second implication is the emergence of such politically charged and value-laden distinctions as natural/social, natural/cultural, natural/artificial and natural/deviant (Rouse 1987, p. 182). The natural immediately falls into a category with the true and the good, while its opposites, the social, the cultural, the artificial and the deviant, fall into a category with the false and the bad. At the same time, the distinctions carry with them the authority and objectivity of natural sciences.

One could also add, that within philosophy there were attempts made to naturalize reason, and hence epistemology, in the 19th century, most famously in the psychologism of John Stuart Mill. This tendency was effectively destroyed by Frege and the linguistic turn of (analytical) philosophy. The drive to naturalize epistemology, the mind and reason itself, returned in Anglo-American philosophy in the 1960s in the writings of philosophers who took their cue variously from psychology, evolutionary theory and neuroscience (see for example Quine 1969; Kornblith 1994). The naturalization fad has also hit the philosophy of science, and the various fields of study concerned with the sciences, in the 1980s, perhaps having its heyday in the early 1990s (see for instance Callebaut 1994).

The biological sciences have a stronger effect on our way of being and self-interpretation than the physical sciences, but the physical sciences can nonetheless be seen as exercising considerable power as the paradigmatic sciences of nature, that is, as setting an example of how to approach nature in order to produce knowledge. As such, the physical sciences are not neutral with regard to human self-understanding. Still, the biological sciences effect a much more direct power on us. One textbook on the philosophy of biology opens with the following words: "The results of the biological sciences are of obvious interest to philosophers because they seem to tell us what we are, how we came to be, and how we relate to the rest of the natural world" (Sterelny and Griffiths 1999, p. 3). The subsection that immediately follows offers a narrower focus on this view:

> What makes someone a human being? The idea that each human being shares with every other human being but with nothing else some essential, human-making features goes back at least to Aristotle [...] Today most people suppose this essence is genetic, and that the job of the Human Genome Project is to reveal the genetic essence of humans. (Sterelny and Griffiths 1999, p. 7)

The effects of the natural sciences on us, how they affect our self-interpretation, what it is to be human, and hence what one can be, are enormous but at the same time subtle and pervasive, in ways I have merely suggested here.

My second example, of how natural sciences can affect our way of being, is more specific and concrete. I take it from Landecker's (1999) discussion of the notorious legal case of Moore v. Regents of the University of California. In the 1980s the word "biological" gained currency as a noun (although its first recorded use as a noun was in 1921 according to the Oxford English Dictionary), to describe living materials produced from organisms. Hence human biologicals are products made from human tissue, blood, etc. They can be of great scientific and economic value. The best known examples are the cell lines Mo and HeLa. HeLa was produced from a tissue sample taken from a cervical lesion of Henrietta Lacks in 1951.[3] Lacks, a black woman, was diagnosed with cancer and died eight months later, but her cells lived on in the HeLa cell line which was mass produced, at first to develop and test the Salk polio vaccine (Landecker 1999, p. 212). The HeLa cell line was a great success and in 1954, three years after it was first developed, a biological supply company marketed the cell line for sale to laboratories.

In 1984 the Regents of the University of California were assigned a patent on the Mo cell line, which had been produced in the late 1970s from the spleen cells of a patient named John Moore. Moore sued the Regents of the University of California for his property rights over the cell line. The California Supreme Court ruled in 1990 that the case was not about a breach of property rights (Landecker 1999, pp. 205–206).

---

[3]For details about the story of Henrietta Lacks, see Skloot's (2010) wonderful book *The Immortal Life of Henrietta Lacks*.

Moore v. the Regents of the University of California was a landmark case, and the ruling of the California Supreme Court has served as a precedent, effectively barring other donors of original cells from participating in patents of their cell lines or profiting from the economic success of their cell lines. The reasoning is that it is not "their" cell lines at all; they merely donated the original cells, based on which cell lines were *invented*. The court reasoned that to

> lend credence to any form of participation in this patent by Moore was ill-advised due to the wider negative consequences this would have in hindering biomedical research by "*restricting access to the necessary raw materials*". (Landecker 1999, p. 206, emphasis added)

In this power struggle between scientists (and not only scientists but also the administrators, managers, lawyers and investors concerned) and bio-sample donors, the human being is a resource for necessary raw materials for the production of biologicals. The human being has become, within this system, a biologicals resource and she must now fight for the ownership and property right over her own body as this resource. It is an irrelevant feature of the example that this particular struggle partly took place in the courts. The point is that a certain scientific practice has turned the human body into a biological resource, giving rise to a struggle for the ownership of this resource. One strategy in this struggle is to start a legal dispute about ownership. Another strategy in this struggle would be to criticize the "discourse of truth" and the techno-scientific practice that has turned the human body into a biological resource.

The two examples above demonstrate how the natural sciences can have a drastic effect on ourselves. The natural sciences exercise great power and non-scientists get caught up in the power struggles no less than the scientists themselves.

3.4 Politics of Truth

In this section I shall first discuss Foucault's use of the phrase "politics of truth" and then the critique that I am proposing under that title. My discussion of Foucault's use of the phrase focuses on two quotes. The first quote is from an interview conducted in 1977 and the second is from a lecture which Foucault gave in 1978.

Foucault's first use of the phrase "politics of truth" I am aware of is in an interview originally published in Italian in 1977, which in English translation is entitled "Truth and Power" (Foucault 2002a [1977]). I shall only be concerned with the last question of the interview, the answer to which Foucault gave in writing. The question posed to Foucault is a familiar one: in the light of his work, what is one to do, and in particular, what is an intellectual to do. Foucault replies by giving a sketch of the history of "the intellectual" as a cultural, social and political critic and conscience. Foucault contrasts the "universal intellectual," typically embodied by the writer of the Marxist tradition, with the "specific intellectual" emerging after the Second World War. The roots of the former go back to Enlightenment writers such as Voltaire. The roots of the latter can be found among the biologists, or more accurately the post-Darwinian evolutionists, of the 19th century and the physicists of early 20th century. The universal intellectual criticizes from a universal viewpoint, in the name of eternal truths and universal values. His standpoint is external to the situation, system or struggle. The specific intellectual is conversely in the midst of the situation. He is typically an expert who is a participant in the struggle and knows it in its local and concrete details. In Foucault's polemic terms, the specific intellectual is "the strategist of life and death," while the universal intellectual is "the rhapsodist of the eternal" (Foucault 2002a [1977]).

It turns out that when it comes to the question of "what to do," Foucault's work presents a task for the specific intellectual but has little use for the universal intellectual. This is not only because Foucault is more interested in specific, local struggles than in global theorizing, nor because the universal intellectual is detached from the situation and so out of touch with it, but because the universal intellectual is nearly extinct. Foucault claims that there are hardly any universal intellectuals around anymore. Many scientists and experts, however, are nowadays forced to become "specific intellectuals" because of the political responsibilities they have due to their position, for instance as nuclear scientists or computer experts (Foucault 2002a [1977]). Scientists and experts are participants in a system of power and knowledge, whether they like it or not. It is because of the present configuration of "the general politics of truth" that the specific intellectual becomes important:

> The important thing here, I believe, is that truth isn't outside power, or lacking in power: contrary to a myth whose history and functions would repay further study, truth isn't the reward of free spirits, the child of protracted solitude, nor the privilege of those who have succeeded in liberating themselves. Truth is a thing of this world: it is produced only by virtue of multiple forms of constraint. And it induces regular effects of power. Each society has its régime of truth, its 'general politics' of truth. (Foucault 2002a [1977])

Foucault uses here the terms "régime of truth," "general politics of truth" and "economy of truth" interchangeably to emphasize that truth is not lofty and pure, but "a thing of this world" with an immense political and economic significance. The politics of truth which Foucault has in mind here have to do with how truth is organized on the global level of society, not on the local level of specific struggles (Foucault's use of the word "truth" may strike some as careless, or worse (Shackel 2005, pp. 299–304), but there are reasons for this use which I shall discuss below). I could use the term "*general* politics of truth" for this idea, but I prefer to use "régime of truth" and reserve "politics of truth" for more local struggles. A régime of truth, then, is constituted by

(1) the types of discourse which it [society] accepts and makes function as true
(2) the mechanisms and instances which enable one to distinguish true and false statements,
(3) the means by which each is sanctioned;
(4) the techniques and procedures accorded value in the acquisition of truth;
(5) the status of those who are charged with saying what counts as true. (Foucault 2002a [1977])

Foucault also lists five characteristics of the *modern* régime of truth. They do not fully match the five constituent parts just quoted, but I shall now briefly discuss these points and match them when possible.

On the first point, in modern Western societies it is above all the scientific discourses which are accepted and made function as true. Whenever there is a dispute about what the truth of a matter is, scientific knowledge trumps any other claim to truth.

Regarding the second point, Foucault does not say what mechanisms are used nowadays to distinguish true and false statements, but these mechanisms are to be found within scientific discourses and the institutions where scientific discourses are produced and transmitted. Their main characteristic is that they are under the control of a "few great political and economic apparatuses" (Foucault 2002a [1977]). Foucault mentions four such apparatuses: university, army, writing and media. We can add the fifth: corporations. Much of research, for instance in biomedical sciences, takes place in corporations, and both the production and the products of that scientific knowledge are very strictly under the control of those corporations. This will become particularly clear in Chap. 5.

As concerns the third point, the sanctioning of truth and falsehood is characterized by the great political and economic incitement to which truth is subjected (Foucault 2002a [1977]). There is a constant demand for truth. Much of this demand is for economic production, that is, demand for knowledge in order to develop new products, to improve the production of goods, to create and manage markets, etc. Much of the demand is also for political reasons, the prime example of which is knowledge tied to biopower, that is, knowledge needed for the management of populations and the disciplining of individuals. Furthermore, knowledge is widely circulated through society and an object of immense consumption, through schools and education, social media, newspapers, magazines, television and so on and so forth.

Foucault is silent on the fourth point, the techniques and procedures presently accorded value in the acquisition of truth. But one is led to think about what ways are considered most reliable, efficient and effective for producing scientific knowledge. This is therefore a question of what is now considered good scientific practice.

On the fifth and last point, regarding the status of those who are charged with saying what counts as true, Foucault must have had in mind the role of scientists and experts in society and the authority that they have. It is one of the characteristics of the modern régime of truth, according to Foucault, that "truth" (or rather scientific knowledge) is constantly a matter of political debate and social confrontation. The scientists and experts tend to get caught up in these debates, where they can then serve the function of specific intellectuals.

Let me now sum up Foucault's description of the five characteristics of the modern régime of truth:

(1) Truth is centred on the form of scientific discourse and the institutions which produce it.
(2) Truth is subject to constant economic and political incitement.
(3) Truth is the object, under diverse forms, of immense diffusion and consumption.

(4) Truth is produced and transmitted under the control, dominant if not exclusive, of a few great political and economic apparatuses.

(5) Truth is the issue of a whole political debate and social confrontation. (Foucault 2002a [1977])

These features can be found in both of the examples of local struggles, of politics of truth, which I discuss in the following two chapters. My discussion there will give more substance and I hope plausibility to Foucault's analysis of the modern régime of truth.

I did promise to come back to the question of "truth." Foucault's use of the word where he talks about truth being a thing of this world may seem questionable, but his use of "truth" in the quote immediately above will surely make some cringe. How can truth be produced and transmitted, diffused and consumed? Analytic philosophers have long ago let the air out of the concept and shown it to be empty and of little use for anything but rhetoric or perhaps semantics. In a way, Foucault's purpose is quite similar: to get rid of the lofty, pure and abstract concept of truth. The difference is that after deflating the concept of truth, Foucault does not leave it as a tool for rhetoric or a technical term in logic and semantics. Instead of being empty and useless, Foucault's concept of truth is concrete and earthly: truth is a thing of this world.

Foucault's "truth" is not just a true sentence. True, in everyday language the words "true" and "truth" are often used simply as shorthand for specific true sentences. But, as at the beginning of the previous sentence, in everyday language the words "true" and "truth" are often also used for emphasis or to convey importance or authority of a specific true sentence or set of true sentences. In that use not all true sentences are worthy of the title "truth." A computer could produce true sentences until it crashes, for instance a sentence like "this sentence was produced at time t" where t is the time at which the sentence is produced. Even if true, these sentences would be of no value or interest—they would not have effects of power even if produced through "multiple forms of constraint." Foucault's truths are important. They are knowledge as opposed to mere information, let alone "true claims."

One could still object that Foucault does *not really* mean truth by "truth," but either what is taken to be true or simply knowledge—or perhaps scientific knowledge. Truth is indeed scientific knowledge in Foucault's analysis of the modern régime of truth, but only because it just so happens that in the modern régime of truth, scientific knowledge plays the role of truth. Still, Foucault claims to mean something quite specific by "truth": "by truth I do not mean 'the ensemble of truths which are to be discovered and accepted', but rather 'the ensemble of rules according to which the true and the false are separated and specific effects of power attached to the true'" (Foucault 2002a [1977]).

Taken literally as a definition of the concept of truth this would be circular and puzzling, and it would reduce his analysis of the modern régime of truth to nonsense. What I take Foucault to be refusing is that "discovery and acceptance" is in any way an important feature of truth. What is important about truth is that it is produced through a mechanism of rules and given great value, authority and power. This would also be the case in a society where the truth is produced through the interpretation of dreams, retelling of myths, and the declarations of village elders. Such truths would also be "produced only by virtue of multiple forms of constraint" and they would surely "induce regular effects of power." And this is also why we cannot replace Foucault's "truth" with "what is taken to be true," or belief: the emphasis is not on acceptance of something as true, but on the mechanism that separates the true and the false and gives the true a certain authority.

We could use "knowledge" for Foucault's "truth," and I think that would be more accurate. The advantage of the term "truth" is that it conveys more authority than "knowledge"; it is strong enough to carry the connotation of production and power that Foucault wants it to have. We can, in any case, use the terms "scientific knowledge" and "scientific facts" for "truth" within the context of the modern régime of truth, because in that context scientific knowledge or facts have the function of truth. This concrete and earthly meaning given to the term "truth" is indeed somewhat unusual. But I think one can only object to that use if one insists either on a metaphysically inflated concept of truth, or the empty, deflated concept of truth of analytic philosophy.

Let me now move on to the second quote from Foucault regarding the politics of truth. It is from a lecture which he gave to the French Society of Philosophy on May 27, 1978. In English translation it is entitled "What is critique?" The lecture was one of a series of lectures about Kant and the Enlightenment, which Foucault gave on various occasions from 1978 to 1984.[4] The lecture's frame of reference was Kant's essay "Was ist Aufklärung?"—What is Enlightenment? Foucault is here primarily concerned with the idea of a critique and the history of the 'critical attitude' in particular in relation to the Enlightenment. The context is closely related to that of the discussion above: there Foucault was asked in effect how one could have a critical attitude given his previous work. He answered by discussing the role of the intellectual in the modern régime of truth. The lecture I discuss here can be seen as a continuation and elaboration of this theme: what does it mean to have a critical attitude, to produce a critique?

Foucault starts his lecture with a discussion of the history of the critical attitude, or what he describes as one possible variation on that history. On Foucault's variation, the history of the critical attitude is closely tied to the history of governance. Foucault identifies the origin of governance in the Christian idea of the individual being lead to salvation through "a total, meticulous, detailed relation of obedience" to someone (Foucault 1997, p. 26). This obedience was related to truth in at least three ways: to truth as dogma, to truth as special knowledge of individuals and to truth as a reflective technique "comprising general rules, particular knowledge, precepts, methods of examination, confessions, interviews etc." (Foucault 1997, p. 26). This governance is individual, spiritual and directed at salvation, but it already involves complex relations of knowledge and power. This model of governance, with similar relations of knowledge and power, proliferated and found new applications in the Renaissance period. Foucault describes the 15th and 16th centuries as the time when

---

[4]"What is Critique" as well as some of the other lectures and a related interview have been published in English translation in a collection with the title *The Politics of Truth* (Foucault 1997).

this kind of governance was taken up in society, in a secular context, and applied to a number of different fields (children, the poor, beggars, the family, the army, cities, states, etc.). Central to this explosion in governance, or what Foucault calls "governmentalization," were two fundamental questions: how to govern and how *not* to be governed. It is in the second question that the critical attitude is born. Foucault states the question as: "how not to be governed *like that*, by that, in the name of those principles, with such and such an objective in mind and by means of such procedures, not like that, not for that, not by them" (Foucault 1997, p. 28). The question is therefore not that of *whether* to be governed, but *how* (not) to be governed. According to Foucault, it is this question of how not to be governed that historically gave rise to the critical attitude. Foucault formulates this as a possible definition of critique: "the art of not being governed quite so much" (Foucault 1997, p. 29).

Foucault goes on to point out that governmentalization, and hence also critique, have at their core the relations between truth, power and the subject—or the relation of the subject to truth (or knowledge) and power. Governmentalization is then the way "individuals are subjugated in the reality of a social practice through mechanisms of power that adhere to a truth" (Foucault 1997, pp. 31–32). This characterization of governmentalization leads Foucault to a revised definition of critique, which is also the second quote on which my discussion of Foucault's "politics of truth" focuses:

> Critique is the movement by which the subject gives himself the right to question truth on its effects of power and question power on its discourses of truth. [...] Critique would essentially insure [sic] the desubjugation of the subject in the context of what we could call, in a word, the politics of truth. (Foucault 1997, p. 32)

Here "politics of truth" is the system of truth and power in which the subject finds him- or herself, the system in which the subject becomes subjugated. One cannot get outside of the politics of truth, but one can resist and avoid subjugation by criticizing the elements of power and truth, that is, by practicing a critique. In modern societies it is scientific knowledge that has the role of truth. Hence a critique would nowadays have the twofold task of questioning science on its effects of power and questioning power on its scientific discourse, with aim of desubjugating the subject. Like Kant in "Was ist Aufklärung?", Foucault is calling for courage: courage to think for yourself. He is opposing the thoughtless following of authority. Foucault's figures of authority are not the priest, the prince and the tax-collector, but the régime of truth within which we are governed. In order to resist or avoid subjugation, we must have a critical attitude to the elements of the modern régime of truth.

So far I have been merely presenting and discussing Foucault's ideas. Now I shall attempt to go beyond exposition and outline a critique for which I use Foucault's term "politics of truth."

If the *régime of truth* is the configuration of power and truth at a given time, then *politics of truth* is the dynamics between the subject and the régime of truth. I use "politics of truth" more specifically to refer to active resistance to certain elements in the régime of truth, in particular when the resistance takes the form of a critique. This critique is directed both at "the discourse of truth," that is, the scientific knowledge production in question, and at "power effects." This resistance is always local. Global resistance is futile since science always has power effects and power, in its modern configuration, must rely on scientific discourse. But this critical analysis may, for instance, show that specific scientific knowledge or a certain scientific programme should not have the sort of power effects it does. In this case one is criticizing a scientific discourse on its effects of power. When criticizing science on its power effects one can ask why it has the power effects it does (and what they are) and whether they are warranted. Criticizing power on its discourse of truth, that is, on its reliance on scientific discourse, can take various forms. One could for instance question the authority of the scientific discourse, or the way specific scientific knowledge is used for the sake of governing or for the sake of having certain effects on individuals (changing what or how they can think or how they can and do behave etc.), or one could even question the epistemological basis of the scientific discourse. Why is the authority of science required? Does the scientific discourse really provide that authority? Who benefits? What are the motives?

Questioning the epistemological basis of a science is a conventional way of criticizing a scientific discourse. It focuses on such issues as the epistemic grounds of the scientific discourse, the methods and procedures used, and whether the science in question is good and proper science. Another way of criticizing science is ethical. It focuses on such issues as the applications of scientific knowledge (is this particular use of scientific knowledge harmful or dangerous?) or whether any harm was or could be done during the production of specific scientific knowledge. In very simplistic terms the first kind of critique could say: "this is bad science!" (an accusation often directed at studies of race and intelligence, for example). And the second kind of critique could say: "this use of science is dangerous/harmful!" (for example, atomic bombs or human genome editing) or "this scientific research is unethical!" (for example, when human research subjects are harmed). When bioethics is critical of science or scientific knowledge, it is almost always concerned with this kind of a critique.

These two kinds of critique are not excluded from the politics of truth. They become special cases. One can, of course, criticize science on the grounds that it is bad science, or criticize the application of science for being harmful or dangerous. A politics-of-truth kind of critique is more importantly concerned with the relations of the subject to power and truth (or power and scientific knowledge in this case). It can criticize the power effects of certain scientific knowledge, even when the science in question is not "bad science" and even when the power effects do not consist in harmful applications of scientific knowledge. And throughout this criticism, one needs to consider the relevant elements of the régime of truth, since they form the basis for the relations of power and truth in question. That is, one must ask what economic and political forces create need and demand for that knowledge, how the knowledge is produced, transmitted, diffused and consumed; and how it gives rise to a political debate or social confrontation. If a given case is a case of bad science or potentially harmful applications of science, that will of course be a central element in the criticism. But, to repeat, it need not be a case of bad science or potentially harmful applications to give rise to a truth-political criticism, since there is no science without power relations and vice versa.

Politics of truth, as a critique, has two objectives: one modest, the other immodest. The modest objective is to arm oneself against the power effects of science. It is to become aware of the subjugating elements of scientific knowledge and find ways of resisting being subjugated. This objective is modest but selfish. One tries to change oneself, and perhaps the reader. It is not about unmasking an ideology or revealing a false consciousness. It is about understanding what is happening in the dynamics between oneself and the system of truth and power, and what one can do to influence these dynamics. Hence the critique aims not merely at understanding relations of power and knowledge but to change them—to the extent that they affect oneself. And the understanding is already the first step in changing them.

The immodest objective is not only to defend oneself but to change the very relations between us and the régime of truth—or even the régime of truth itself. One may try to change the way people think about a given scientific discourse, in order to reduce or reshape the power effects it has. Or one may try to change the way of thinking within science—to change how scientists think about their subjects, or to change the style of thought or the very form of a specific science. One may also try to change the entire structure of the régime of truth, which is what Foucault once suggested:

> The essential political problem for the intellectual is not to criticise the ideological contents supposedly linked to science, or to ensure that his own scientific practice is accompanied by a correct ideology, but that of ascertaining the possibility of constituting a new politics of truth. The problem is not changing people's consciousness—or what's in their heads—but the political, economic, institutional régime of the production of truth. (Foucault 2002a [1977])

I have considerable doubts that a critical analysis of a scientific knowledge production could achieve a transformation of the régime of truth, but I do not exclude the possibility. Hence an immodest objective.

The critique I have described has both historical and philosophical aspects. On one hand it presents a history of the scientific knowledge production in question. This history takes its cue from Foucault's archaeology. It does not necessarily present the epistemic conditions for the scientific knowledge, but it does endeavour to show the historical contingency of scientific knowledge. The point is not to demonstrate that scientific knowledge or its objects are a "mere social construct," but rather how the particular scientific knowledge is contingent and earthly and for what purpose it is made. Only showing that certain scientific knowledge classification or perspective is not inevitable, can be empowering.[5] Still, the point is not to undermine the authority of all science, but to question the authority of specific scientific knowledge through a historical account of it in order to better resist its subjugating power effects. This history is therefore not written for its own sake, it is written to change how we look at certain relations in the present: it is a history of the present.

On the other hand the critique is philosophical, as it presents an analysis of the relations between the subject, knowledge and power. Studying relations between the subject and knowledge places one in the familiar field of epistemology. Studying

---

[5]Hacking (1999) has made this point nicely in his book *The Social Construction of What?*, see Chap. 1 "Why ask what?".

relations between the subject and power places one in the familiar fields of ethics. Studying relations between the subject and the régime of truth, that is, the subject's relation to the present configuration of power and knowledge, places one in an unfamiliar field where ethics and epistemology merge. This critique, that is engaged in politics of truth, offers an alternative to the two conventional critiques of science (epistemological and ethical). It does not exclude epistemological or ethical critique but makes use of them when they are relevant.

To my knowledge, there are no studies that approach scientific knowledge production in exactly the way I have here described, but a lot of work has been done that would fall in one way or another under the term "politics of truth." Trivially, all critique of science is engaged in a politics of truth, from Karl Popper to critics within the field of science studies. Foucault's own historical studies are the best examples. Foucault was apparently surprised that many considered his *Madness and Civilization* an anti-psychiatry book, but he was more consciously and explicitly critical in his power/knowledge studies: *Discipline and Punish* and *The History of Sexuality*, which I discussed in the previous sections. Another book that I discussed above directly addresses issues of power and knowledge in relation to current knowledge production, namely Rouse's (1987) *Knowledge and Power*. But Rouse's work, under a strong influence from Heidegger, puts great emphasis on issues of interpretation and scientific practice, at the cost of historical concerns, and it stresses global concerns about the nature of science rather than local concerns about specific cases of science.

Two books by Ian Hacking do, however, stress the historical aspect of specific cases of scientific knowledge: *Rewriting the Soul* and *Mad Travellers*. Although they are not directly analyses of power/knowledge relations, they do tell the story of the interaction between certain scientific knowledge/practice and its subjects. Their subject-matter is multiple personality disorder (now classified as dissociative disorder) and compulsive travelling (now non-existent), respectively. Their effects, and I must assume that these effects are intentional, are to change the way one sees the relations between psychiatry and its subjects, and undermine the authority of psychiatry—precisely by showing the historical contingency of psychiatric knowledge, classifications and objects. Hacking is rather nominalist about psychiatric classifications, but without the implication that the suffering of the mentally ill is in any way not real. Hacking's two historical studies differ from the critique that I am proposing, as they do not directly or explicitly analyse power/knowledge relations, nor in any detailed way the mechanisms of producing, circulating, diffusing and consuming the scientific knowledge in question.

There is a great deal of science studies literature that is critical of science, often in ways inspired by Foucault or otherwise similar to the approach I am suggesting here (e.g., Rose 2009). But I am not aware of any that directly and explicitly takes up Foucault's idea of a critique as I have presented it here.

One obvious place to look for discussions on power, knowledge and the politics of truth is studies on gender, race and sexuality. I avoid these fields of study for two reasons. First, the terrain has been thoroughly worked: the paths have been laid, borders marked, trenches dug. Entering these vast and complex fields would distract too much from the work I am trying to do. I hope to keep a better focus by turning my attention to more obscure research subjects and less obviously subjugated subjects than women, blacks, and homosexuals—namely left-handers and Icelanders. Second, in the spirit of Foucault's specific intellectuals, I am concerned with issues that I know about and care about because of my own personal involvement with them.

In this chapter I have discussed the relation between power and knowledge in Foucault's work. After discussing the power dynamics of scientific knowledge about humans, the so-called human sciences, I considered the value of Foucault's concept of "biopower" for power/knowledge analysis of scientific knowledge. I then argued that Foucault's power/knowledge analysis could also apply to the natural sciences, in particular to the life sciences when their research subject is man. I ended my discussion with an account of Foucault's idea of a critique in the context of politics of truth, arguing that it offers an alternative to the two conventional types of critique of science (bad in its methodology or harmful in its consequences). In the two remaining chapters of this thesis I will discuss two cases of knowledge production. My approach to the cases will be based on the ideas about power and knowledge discussed in this chapter and on the ideas about the historical conditions for knowledge explored in the previous chapter. These ideas do not constitute a methodology, still less "theory," but they indicate an approach and an attitude to scientific knowledge which is critical without being anti-science.

References

Achbar, Mark (ed.). 1994. *Manufacturing Consent: Noam Chomsky and the Media*. Montréal: Black Rose Books.

Callebaut, Werner (ed.). 1994. *Taking the Naturalistic Turn, or How Real Philosophy of Science is Done*. Chicago: University of Chicago Press.

Esposito, Roberto. 2008. *Bíos: Biopolitics and Philosophy*. Minneapolis: University of Minnesota Press.

Foucault, Michel. 1972. *The Archaeology of Knowledge, and the Discourse on Language*. New York: Pantheon. Trans. A. Sheridan Smith. Originally *L'archéologie du savoir*. Paris: Gallimard, 1969.

Foucault, Michel (ed.). 1975. *I, Pierre Rivière, Having Slaughtered My Mother, My Sister, and My Brother*. New York: Pantheon Books. Originally *Moi, Pierre Revière, ayant égorgé ma mère, ma soeur et mon frère…* Paris: Gallimard, 1973.

Foucault, Michel. 1977a [1971]. Revolutionary Action: 'Until Now'. In *Language, Counter-Memory, Practice: Selected Essays and Interviews by Michel Foucault*, ed. Donald F. Bouchard, 218–233. Ithaca: Cornell University Press. This discussion was originally published in *Actuel* in 1971.

Foucault, Michel. 1977b. *Discipline and Punish*. New York: Pantheon. Trans. A. Sheridan. Originally *Surveiller et punir: Naissance de la prison*. Paris: Gallimard, 1975.

Foucault, Michel. 1978. *The History of Sexuality I: An Introduction*. New York: Pantheon. Trans. R. Hurley. Originally *Histoire de la sexualité I: La volonté de savoir*. Paris: Gallimard, 1976.

Foucault, Michel. 1980a [1976]. Two Lectures. In *Power/Knowledge: Selected Interviews and Other Writings 1972–1977*, ed. Colin Gordon, 78–108. New York: Pantheon. These two lectures were given in 1976, but first appeared in Italian translation in Michel Foucault, *Microfisica del Potere* (Turin, 1977).

Foucault, Michel. 1980b [1972]. On Popular Justice: A Discussion with Maoists. In *Power/Knowledge: Selected Interviews and Other Writings 1972–1977*, ed. Colin Gordon, 1–36. New York: Pantheon. This discussion was originally published in *Les temps moderne* in 1972.

Foucault, Michel. 1980c [1977]. The Confession of the Flesh. In *Power/Knowledge: Selected Interviews and Other Writings 1972–1977*, ed. Colin Gordon, 194–228. This conversation was originally published as "Le jeu de Michel Foucault" in *Ornicar?* in 1977.

Foucault, Michel (ed.). 1980d. *Herculine Barbin: Being the Recently Discovered Memoirs of a Nineteenth-Century French Hermaphrodite*. New York: Pantheon Books.

Foucault, Michel. 1980e [1976]. Questions on Geography. In *Power/Knowledge: Selected Interviews and Other Writings 1972–1977*, ed. Colin Gordon, 63–77. This interview first appeared as "Questions à Michel Foucault sur la géographie," in *Hérodote* 1 (1976).

Foucault, Michel. 1988 [1984]. On Power. In *Politics, Philosophy, Culture: Interviews and other Writings 1977–1984*, ed. Lawrence D. Kritzman, 96–109. New York and London: Routledge. This interview was conducted in 1978, but appeared first, and then only in part, in *L'express* in 1984.

Foucault, Michel. 1989a [1971]. Rituals of Exclusion. In *Foucault Live: Interviews, 1966–84*, ed. Sylvère Lotringer, 63–72. New York: Semiotext(e). This interview was originally published in *Partisan Review* in 1971.

Foucault, Michel. 1989b [1984]. The Concern for Truth. In *Foucault Live: Interviews, 1966–84*, ed. Sylvère Lotringer, 293–308. New York: Semiotext(e). Trans. J. Johnston. This interview appeared first in *Le Magazine littéraire*, May 1984.

Foucault, Michel. 1989c [1976]. I, Pierre Rivière. In *Foucault Live: Interviews, 1966–84*, ed. Sylvère Lotringer, 131–136. New York: Semiotext(e). Trans. J. Johnston. This interview first appeared in *Cahiers du cinéma* in November 1976.

Foucault, Michel. 1997. What is Critique? In *The Politics of Truth*, ed. Sylvère Lotringer, 23–82. New York: Semiotext(e). This is the text of a lecture given on May 27, 1978. First published in the *Bulletin de la Société française de philosophie*, t. LXXXIV, 1990.

Foucault, Michel. 2000 [1971]. Nietzsche, Genealogy, History. In *Essential Works Volume 2: Aesthetics, Method, and Epistemology*, ed. James Faubion and Paul Rabinow, 369–391. London Penguin Books. This essay first appeared in *Hommage à Jean Hyppolite*. Paris: Presses Universitaires de France, 1971.

Foucault, Michel. 2002a [1977]. Truth and Power. In *Essential Works Volume 2: Aesthetics, Method, and Epistemology*, ed. James Faubion and Paul Rabinow, 111–133. London: Penguin Books. This interview first appeared in Italian translation as "Intervista a Michel Foucault," in Michel Foucault, *Microfisica del Potere* (Turin, 1977).

Foucault, Michel. 2002b [1978]. About the Concept of the "Dangerous Individual" in Nineteenth-Century Legal Psychiatry. In *Essential Works, Volume 3: Power*, ed. James Faubion and Paul Rabinow, 176–200. London: Penguin Books. First published in English in the *International Journal of Law and Psychiatry* in 1978.

Foucault, Michel. 2003 [1976]. *"Society Must Be Defended": Lectures at the Collège de France 1975–1976*, ed. Arnold Davidson, Mauro Bertani, Alessandro Fontana, and François Edwald. New York: Picador.

Foucault, Michel. 2007 [1978]. *Security, Territory, Population: Lectures at the Collège de France 1977–1978*, ed. Arnold Davidson, Michel Senellart, and François Edwald. New York: Picador.

Foucault, Michel. 2008 [1979]. *The Birth of Biopolitics: Lectures at the Collège de France 1978–1979*, ed. Arnold Davidson, Michel Senellart, François Edwald, and Alessandro Fontana. New York: Picador.

Foucault, Michel. 2014 [1980]. *On the Government of the Living: Lectures at the Collège de France 1979–1980*, ed. Arnold Davidson, Michel Senellart, François Edwald, and Alessandro Fontana. New York: Picador.

Foucault, Michel, and Gilles Deleuze. 1977 [1972]. Intellectuals and Power: A Conversation Between Michel Foucault and Gilles Deleuze. In *Language, Counter-Memory, Practice: Selected Essays and Interviews by Michel Foucault*, ed. Donald F. Bouchard, 205–217. Ithaca: Cornell University Press.

Hacking, Ian. 1986. Making Up People. In *Reconstructing Individualism*, ed. Thomas C. Heller, Morton Sosna, and David E. Wellbery, 222–236. Stanford, CA: Stanford University Press; republished in Ian Hacking. 2002. *Historical Ontology*. London and Cambridge, MA: Harvard University Press.

Hacking, Ian. 1992. 'Style' for Historians and Philosophers. *Studies in History and Philosophy of Science Part A* 23 (1): 1–20.

Hacking, Ian. 1999. *The Social Construction of What?*. Cambridge, MA: Harvard University Press.

Hacking, Ian. 2007. Kinds of People: Moving Targets. *Proceedings of the British Academy* 151: 285–318.

Koopman, Colin. 2013. *Genealogy as Critique: Foucault and the Problems of Modernity*. Bloomington, IN: Indiana University Press.

Kornblith, Hilary (ed.). 1994. *Naturalizing Epistemology*, 2nd ed. Cambridge, MA, and London: The MIT Press.

Landecker, Hannah. 1999. Between Beneficence and Chattel: The Human Biological in Law and Science. *Science in Context* 12: 203–225.

May, Todd. 1993. *Between Genealogy and Epistemology: Psychology, Politics, and Knowledge in the Thought of Michel Foucault*. University Park, PA: The Pennsylvania State University Press.

Miller, James. 1994. *The Passion of Michel Foucault*. New York: Doubleday.

Pickering, Andrew. 1984. *Constructing Quarks: A Sociological History of Particle Physics*. Edinburgh: Edinburgh University Press.

Quine, Willard Van Orman. 1969. Epistemology Naturalized. In *Ontological Relativity & Other Essays*, 69–90. New York: Columbia University Press.

Rabinow, Paul, and Nikolas Rose. 2006. Biopower Today. *BioSocieties* 1 (2): 195–217.

Rose, Nikolas. 2009. *The Politics of Life Itself: Biomedicine, Power, and Subjectivity in the Twenty-First Century*. Princeton, NJ: Princeton University Press.

Rouse, Joseph. 1987. *Knowledge and Power: Toward a Political Philosophy of Science*. Ithaca and London: Cornell University Press.

Shackel, Nicholas. 2005. The Vacuity of Postmodernist Methodology. *Metaphilosophy* 36: 295–320.

Skloot, Rebecca. 2010. *The Immortal Life of Henrietta Lacks*. New York: Random House.

Sterelny, Kim, and Paul E. Griffiths. 1999. *Sex and Death: An Introduction to Philosophy of Biology*. Chicago and London: The University of Chicago Press.

Chapter 4
Left-Handers as Subjects of Science

In this chapter and the next I engage in two case studies. The first concerns left-handed people and studies aimed at answering the question of whether left-handers have reduced longevity compared to right-handers.[1] In the first section I shall begin by briefly discussing the history of left-handedness and, in the second section, how this phenomenon emerged as an object of scientific study. In the third section, I shall discuss in some detail three studies on the longevity of left-handers. In the fourth and last section I discuss the production of knowledge about left-handers in the context of the modern régime of truth. The aim of this chapter is

(1) to draw a picture of a piece of scientific research which illustrates how pre-scientific ideas and values can influence the direction and very content of scientific research;
(2) to analyse the relations of power which appear on the one hand in the rise of the scientific study of left-handers, and on the other hand in the reactions—the resistance—to the research;
(3) to identify the features of the modern régime of truth apparent in the scientific discourse on left-handers; and
(4) to criticize the scientific study of left-handers.

The first aim is not exactly an example of Foucauldian archaeology, as I am not identifying any epistemic organizing principles underlying the scientific discipline within which this scientific research is conducted. The idea is rather that less grand things than epistemes can also affect the production and even content of scientific knowledge, such as the history of the concepts and values that are still at play in the research. This idea may be closer to Ludwik Fleck than Michel Foucault and one of its main aspects has been nicely expressed by Kierkegaard:

---

[1] A version of this chapter has appeared as "Biopolitics and the Longevity of Left-Handers" (Árnason 2017), where I discuss the research on left-handers and longevity in a slightly different context.

© The Author(s) 2018
G. Árnason, *Foucault and the Human Subject of Science*, SpringerBriefs in Ethics,
https://doi.org/10.1007/978-3-030-02813-8_4

> Concepts, like individuals, have their histories and are just as incapable of withstanding the ravages of time as are individuals. But in and through all this they retain a kind of homesickness for the scenes of their childhood. (Kierkegaard 1965 [1841], p. 47)

Nonetheless, this historical concern is closely related to that of Foucault's archae-ologies, in particular that there are historical conditions and constraints necessary for the production of scientific knowledge. Some of these conditions are of the deep structure sort, others are mere surface concepts and values, longing for "the scenes of their childhood." In less poetic terms, the concepts still carry with them traces of functions, connotations and values of their past. Another aspect of the historical part of my study is perhaps closer to Foucault's archaeologies: I endeavour to demon-strate the arbitrariness of the scientific concept of left-handedness and the historical contingency of the scientific knowledge, as it is, about left-handers.

The remaining three aims refer quite straight-forwardly back to my discussion of knowledge, power and the politics of truth in the previous chapter.

4.1   Pre-scientific Conceptions of Left-Handedness

One can only speculate about when left-handedness emerged as a noticeable phe-nomenon, but it is most likely as old as our species. It is quite possible that handedness evolved in close relation to the human brain as the human species emerged.[2] Left-handedness must have become all the more visible as tools became more complex, and as cultures and traditions developed around the use of tools, in particular when the need arose to teach how to use tools in the *right* way. *Homo habilis* is thought to have manufactured the first stone tools as long as 2.5 million years ago, but the first tools complex enough to require instruction to manufacture (and probably to use as well) appeared with *Homo erectus* about 1.6 million years ago (Corballis 1991, pp. 61–62).

It is not known when, how or why asymmetrical handedness evolved in humans, although there is no lack of theories. Michael C. Corballis suggests for instance that handedness may have evolved along with the manufacturing of tools (Corballis 1991, p. 104; see also Uomini 2009), or that it is, more generally, a by-product of "the left-hemispheric control of praxis" where *praxis* is "the organization of pur-poseful, sequential actions in which spatial constraints imposed by the environment are minimal" (Corballis 1991, pp. 213–214).

As far as written records go, it has been long noted that, although most people prefer using their right-hand for most tasks, some people would always rather use their left-hand. Left-handers typically aroused curiosity and were recognized by others as a strange and perhaps evil kind of people. Left has traditionally been associated with everything wrong and bad, and right with everything "right" and good. Left-handers are often called "sinistrals" in studies, *sinister* being the Latin word for "left." The

---

[2]Wilson (1998) argues exactly that in his book *The Hand: How Its Use Shapes the Brain, Language, and Human Culture*; see also Frayer et al. (2012).

Latin word for "right" is *dexter*, and in English people are said to be *dexterous*. From French we have *gauche* (left) meaning awkward or inappropriate. In 1946, psychiatrist Abram Blau noted "in language, negativism is quite recognizable in the connotations of contrariness associated with the various words for the left side, such as gauche, sinister, left-handed compliment, left (radical) party, etc." (Blau 1946, p. 91).

Blau suggests that the word "left" in the political context is negative. The political "left" has its origin in the French National Assembly, which in 1789 decided to sit radicals on the president's left-hand side, the conservative nobles on his right and the moderates in the middle. More recently, psychologist Chris McManus has suggested, like Blau, that "left" in politics is a negative term, but on the grounds that it is one member of the pair right-left, where right is symbolically the positive term and the norm, while left is the negative term and the deviation. McManus then argues that since people, when asked about their views on anything, are more likely to answer with the positive term than the negative term, when an answer requires one or the other (and everything else being equal), people are more likely to claim to be on the right wing in politics than the left. Hence the left will always be in the minority and the right in the majority (McManus 2002, pp. 260–264). Here the negative connotations of the word "left" are claimed to have a direct party-political consequences.

We know that negative associations with left-handedness are strong, at least in the English speaking world, but that pales next to some cultures of former times. In the Pythagorean Table of Opposites, which Aristotle describes in *Metaphysics* (986a25), the right is associated with the limited, the odd, the one, the male, the state of rest, the straight, the light, the good and the square, whereas the left is associated with the unlimited, the even, plurality, the female, the moving, the curved, darkness, the bad and the oblong (see Corballis 1983, pp. 1–2; also Wilbur and Allen 1979, p. 86; and Lloyd 1962). Corballis writes:

> Among the Nyoro of East Africa, for instance, properties associated with the right include man, brewing, health, wealth, fertility, life, the even, the hard, the moon, fidelity, and cattle, whereas the opposing properties associated with the left are woman, cooking, sickness, poverty, barrenness, death, the odd, the soft, the sun, and chickens or sheep (Needham 1967). Essentially the same associations are found among the Gogo of Tanzania. (Rigby 1966; Corballis 1983, p. 2)

Anthropologists have found similar negative associations with the term "left" in other cultures, such as the Nuer of South Sudan and the Temne of Sierra Leone (Dawson 1977, p. 425); the Maori of New Zealand and the Wulwanga of Australia (Hertz 2013 [1909], pp. 343–345), to name just a few. The anthropologist Robert Hertz summarized the view of the left hand in different cultures thus: "To the right hand go honours, flattering designations, prerogatives: it acts, orders, and takes. The left hand, on the contrary, is despised and reduced to the role of a humble auxiliary: by itself it can do nothing; it helps, it supports, it holds" (Hertz 2013 [1909], p. 335).

Whether one is in North-America, among the Gogo of Tanzania or the ancient Greeks, everywhere left-handers seem to be associated with bad things. In my Greek lexicon I find that the Greek word for "left," *aristeros*, can mean "ill-boding" or "ominous" and the word for "right," *dexios*, can mean "boding good" or "fortunate."

My lexicon says, as an explanation for these meanings, that this "sense came from the Greek augurs looking to the North, so that *lucky* omens, which came from the East, were *on the right*, while the *unlucky* ones from the West were *on the left*" (Liddell and Scott 1992).

It is interesting, in comparison, that the Maoris of New Zealand observed tremors in sleeping people and believed that the body had been seized by a spirit—if the tremor occurred on the right side, it foretold good fortune and life, whereas tremor on the left side meant ill fortune and even death (Hertz 2013 [1909], p. 344). Similarly, the native people of Morocco interpreted a twitching of the right eyelid as signifying the return of a member of the family, or other good news, whereas twitching of the left eyelid was a warning of impending death in the family (Corballis 1983, p. 1). Poets seem to agree that bad signs appear to the left, and the good ones to the right. In 270 B.C. the Greek poet Theocritus wrote "my right eye itches now, and shall I see my love?" and in the 18th century John Gay included this bit of verse in one of his fables: "That raven on your left-hand oak/(Curse his ill-betiding croak)/Bodes me no good" (Coren 1992, p. 11).[3]

Religions, too, reflect the negative associations of left-handedness. Buddha sends his adherents down the right-hand path to Nirvana. In the Bible we are told that when God divided the sheep and the goats, he consigned the goats to the left saying "Depart from me, ye cursed." The devil sits to God's left and is often portrayed as left-handed (Shute 1994, pp. 131–132).

An odd consequence of the negative view of left-handedness is that left-handers are sometimes believed to have special gifts. Nowadays there is rarely a popular discussion of left-handedness without long lists of famous left-handers. This phenomenon of a positive (but freakish) manifestation of something gone wrong is not unique: there are countless stories of people with mental disabilities, for instance autistic people, who have almost super-human cognitive abilities usually related to numbers or memory (note for instance the film "Rain Man" featuring Dustin Hoffman). The difference is that the talented left-handers enjoy great success: as politicians [former U.S. presidents Gerald Ford, Ronald Reagan, George Bush (sr.), Bill Clinton and Barack Obama (Binks 2008)], artists (notable examples: Michelangelo, Raphael and Leonardo da Vinci), conquerors (Alexander the Great, Charlemagne and Napoleon Bonaparte), classical composers (C.P.E. Bach (son of Johann Sebastian), Beethoven, Mozart, Ravel, Schumann, Paganini), Beatles (Paul McCartney and Ringo Starr (who incidentally have survived their right-handed fellow Beatles)), scientists (Isaac Newton, Albert Einstein, Marie Curie), Hollywood stars (Robert de Niro, Julia Roberts, Tom Cruise, Nicole Kidman) and the list goes on and on.

These lists do not of course show that left-handers are more successful than right-handers. Assuming that roughly 10% of humans are left-handed, such lists could quite easily be results of chance: there are so many people who have some claim to fame that any 10% subset of them chosen at random will form a remarkable list of

---

[3]The quote from Theocritus can be found in Cholmeley (1919), 3rd Idyllium; and the quote from John Gay can be found in Gay (2005 [1727]), p. 97 (Fable XXXVII: The Farmer's Wife and the Raven).

famous people. The point is rather that there is a need to compile such lists and there is a common impression that left-handers are more likely to be unusually intelligent or creative. This does not mean that there is a conflicting idea of the left-handed, on the one hand abnormal and bad, on the other hand superior to the rest. This is rather another manifestation of the perceived abnormality of left-handers: their very talents are taken to be a result of something gone wrong.

Left-handedness has had its place in the dualism manifested in the endless double-column of good and bad, but through the explosion of statistics and statistical reasoning in the early 19th century, left-handedness became bad in a new way: as a deviation from a norm, as an abnormality. The statistical language of norms and deviations does not, at the mathematical level, entail any value judgments, but it evolved as a way of organizing knowledge about human affairs and quickly, if not immediately, found its place in the double-column of good and bad. In the guise of abnormality, left-handedness was taking its first steps towards science.

The concept of left-handedness carries a heavy historical baggage of negative associations and values, even today in our Western liberal and pluralistic societies. Perhaps, after it has wandered a long way to the tidy grounds of modern science, it still secretly longs for the scenes of its childhood.

4.2 The Rise of the Scientific Study of Left-Handers

One consequence of the negative attitudes towards left-handedness is that left-handers have frequently been forced by parents and teachers to use their right-hand, specifically for eating and writing. Until recently, children in the Western world were chastised if they held tools such as a pen or a fork in their left-hand. At school, left-handed children would risk having their left-hand smacked or even tied up if they were found holding the pen in their left-hand. Educators discussed how to solve the problem of stubborn children who would not use the right-hand like everyone else. In 1914 the *Teacher* published an article titled "What shall we do with left-handed pupils?" In the article A. B. Poland, school superintendent in Newark, New Jersey, is quoted saying: "I have not adequate knowledge derived from experience to warrant me in expressing an opinion as to the best means of dealing with left-handed pupils" (McMullin 1914, quoted in Harris 1980, p. 3).

Left-handers, or at least left-handed pupils, were a group that needed to be dealt with, they had to be controlled, managed, disciplined. In order to control, manage and discipline them more effectively, it became important, as the superintendent noted, to gain some knowledge of this sort of people. In the last section I mentioned the psychiatrist Abram Blau. He was a psychiatrist at the New York University College of Medicine and before that a school psychiatrist at the Bureau of Child Guidance, Board of Education in New York city. In 1946 he wrote:

A principal of an elementary public school in one of the better neighbourhoods in New York City remarked to me about the alarming increase of left-handedness among his pupils. This was undoubtedly due to the increased cultural tolerance of sinistral tendencies which should really be discouraged and the child trained to adjust himself normally to this right-handed world. (Blau 1946, p. 90)

Abram Blau set out to acquire the knowledge of left-handedness, so these troublesome pupils could be dealt with properly. His findings were quite what he had expected: in his research monograph *The Master Hand* he claims:

Sinistrality is the product of a contrary attitude on the part of the infant and young child. In other words, sinistrality is thus a symptom or manifestation of an attitude of opposition or negativism along with such other signs as disobedience, refusal to eat, temper tantrums, rebelliousness, etc. In place of a wish to comply with the social and cultural pressures toward the use of the right hand, there exists an active attitude of opposition which manifests itself in the development of sinistrality. (Blau 1946, p. 91)

This was neither new nor unique in academic writings in the early twentieth century. A few years before, the British educational psychologist Sir Cyril Burt had written in his book *The Backward Child* that left-handers were wilful: "Even left-handed girls [...] often possess a strong, self-willed and almost masculine disposition" (Burt 1932, quoted in Corballis 1991, p. 91). In other ways he did not have a very favourable view of left-handers:

They squint, they stammer, they shuffle and shamble, they flounder about like seals out of water. Awkward in the house, and clumsy in their games, they are fumblers and bunglers at whatever they do. (Burt 1937, quoted in Corballis 1991, p. 91)

There was clearly a strong need for knowledge to manage left-handers as a problematic group of people, a need which seems to have given rise to scientific research on left-handedness. We can see here the emergence of power/knowledge relations: left-handers have been singled out as a group that needs to be managed and disciplined with a regard to the truth about left-handers. The demand for knowledge about problematic left-handers results in knowledge that presents them as problematic, enforcing in turn the need to deal with them. This further illustrates two characteristics of the modern régime of truth discussed in Sect. 3.4. First, science is the supreme source of truth. The truth about left-handers must be sought from science, although we all have personal experience of left-handers. For instance, despite all his experience with left-handed pupils, the New Jersey superintendent quoted above noted: "I have not adequate knowledge *derived from experience* to warrant me in expressing an opinion as to the best means of dealing with left-handed pupils" (my emphasis). What is needed is *scientific* knowledge. Second, truth is subjected to great political and economic incitement. Left-handers became a group in education politics, a group that teachers and parents had to deal with, and this required increased knowledge of this group.

Nowadays the incitement for knowledge production about left-handers is more economic than political. The political need to deal with left-handers has drastically decreased in Western schools; at least there is no explicit demand for knowledge about left-handers for the purpose of controlling them and converting them to right-handedness. But there is still a need to manage them, as is borne out by recent books bearing titles such as *Your Left-Handed Child: Making Things Easy for Left-Handers in a Right-Handed World* and *Loving Lefties: How to Raise Your Left-Handed Child in a Right-Handed World*, not to mention books teaching left-handed pupils to write properly, such as *Cursive Writing Skills for Left-Handed Students*, *Left Hand Writing Skills* (three books), and *Handwriting Program for Cursive Left Hand* (*Preventing Academic Failure*). I suspect that this change of educational policy is not only stemming from increased tolerance and a more positive view of left-handers, but even more from the fact that left-handedness is not easily "rectified" without applying certain physical force and even physical constraint over an extended period of time—and such methods are out of fashion. Although educators are nowadays not willing, or even able within the law, to constrain pupils' left hand in order to make them right-handed, there are still people who promote making left-handed children "normal", without the application of physical force or constraint. Just to give one example, a popular German advice book for parents of young children suggests that left-handed children should be "given the opportunity to learn to write with the right hand," not through coercion but "with patience and care" so that the child "is proud to be able to write 'like the others'" (Göbel and Glöckler 2015, pp. 541–549, my translation).

Many of us, especially us left-handers, think it is obviously wrong to "convert" left-handers to right-handedness. Let us, then, imagine a parallel universe, where left-handedness can be not only detected in a foetus (Parma et al. 2017) but also "rectified" with a quick and painless medical procedure. Parents would have the autonomous choice (of course) of converting their left-handed children, after being told that the benefit for the child would be reduced risk of accidents, disease and physical abnormalities and possibly increased life expectancy, in addition to functioning better in our right-handed world. Would any parent say no to that? I do not mean to advocate any *right* to remain a left-hander. My point is simply that (potential) left-handers would most likely be converted to right-handedness if an acceptable method existed.

The school masters and superintendents may not be demanding knowledge about left-handers to be able to change their behaviour, but there is still a need to deal with left-handers, particularly left-handed pupils but also left-handed employees, athletes, musicians, cooks, gardeners and left-handed consumers in general, and this need to deal with them also creates a need for knowledge about left-handedness. That there is an economic incitement for the truth about left-handers is manifested by the great number of books aimed at left-handers and at parents of left-handed children, as well as the proliferation of stores, both virtual and physical, selling products for left-handers. At the time of writing, the online retailer Amazon lists almost 900 books with the word "left-handed" in the title. Not only left-handers themselves are interested in the truth about left-handers; parents, teachers, physicians, employers, designers, manufacturers, and service providers have to deal with left-handedness and left-handers, and this creates a need for knowledge.

Although the need for knowledge about left-handers has changed, the scientific research on left-handedness has probably never been more productive than during the last three decades. It is difficult to explain this scientific interest in left-handedness without referring to the need to "deal with" left-handedness, and the political and economic incitement to which knowledge about left-handers is subjected. Another motivation for left-handers research is the historical baggage of the concept: left-handedness is interesting because it is deviant, abnormal and, after all, perhaps not quite right.

Many scientists studying left-handedness acknowledge that handedness is just a part of a larger phenomenon, namely the laterality of the body, but they have nonetheless a very strong tendency to talk about their studies and results as if the subject is strictly left-handers, or they even explicitly limit their study on human laterality to left-handedness. Most of the research on laterality is concerned with *deviations* from the usual right shift in human laterality (the tendency of humans to be right-sided). When classifying the subjects, questionnaires are used which have mostly, or even exclusively, items regarding hand preference. Much more research is done on left-handedness in particular than the common overall right-sidedness or the lateral asymmetry of the brain and the body. Much of this research aims at establishing statistical links between left-handedness and anything human, despite the fact that not much is known about the causes of left-handedness, nor right-handedness for that matter, and thus no one knows what material relations could underlie the statistical links. All this interest and effort betrays, I believe, the effects of the pre-scientific concept of left-handedness.

Let us now pause and take a brief look at how scientific discourse tries to get a grip on left-handedness. How do the sciences define handedness? First, there are a few quite straightforward observations to be made. To begin with, there is more to handedness than simply "right or left"; for as long as there have been left-handers and right-handers, there have also been the ambidextrous. Still, three types of handedness, rather than two types, does not sufficiently reflect the nature of handedness. It was noted already in a 1927 study that "dextrality and sinistrality are not opposed alternatives, but quantities capable of taking values of continuous intensity and passing one into the other" (Woo and Pearson 1927, p. 199). In addition to that, some people prefer the right hand for some tasks and the left for other tasks in a constant manner. Such a person has a strong hand preference, but might get the same "handedness quotient" on a test as someone who has a weak hand preference, while their hand preferences actually differ wildly. One might therefore talk of a multi-dimensional model of handedness being more appropriate than the binary, trinal or even one-dimensional model (left-right graded scale).

My second observation is that the hands are not the only lateralized part of the human body. Most people have a preference for foot, eye and ear, and this preference may or may not coincide with hand preference. About 10% of humans are left-handed, but approximately 20% are left-footed, 30% "left-eyed" and 40% "left-eared" (McManus 2002, pp. 153–154; see also Porac and Coren 1981). Some people have consistent side dominance, and this is more common among right-handers than left-handers, but many are cross-lateral.

Furthermore, a distinction is to be made between preference and ability. Although this is rare, some people do not prefer the hand which has more strength or better

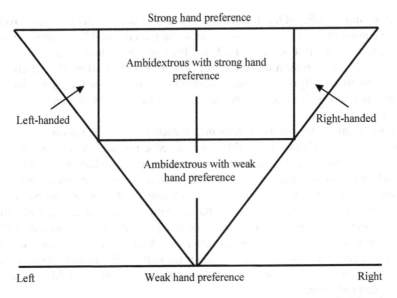

Fig. 4.1 Possible outcomes from tests of hand preference that include both a left-right axis and a strong-weak axis. This figure suggests one way of dividing the field of possible outcomes in order to define left-handedness

motor skills. This applies in particular to people who have little or no hand preference despite one hand being stronger or more skilled than the other. However, the general correlation of hand preference and hand proficiency in non-right-handers has long been a matter of debate (see Bishop 1989; Corey et al. 2001; Brown et al. 2004).

What these observations amount to is that any attempt to make the common sense (or pre-scientific) conception of left-handedness rigorous results in a diffuse picture of the laterality of the brain and the rest of the human body where left-handedness does not stand out in any way. This leaves one with two options for defining left-handedness: One is to make a detailed picture of physical laterality, then take out only what applies to handedness, and assign specific cut-off values for hand preference (for both the strong-weak axis and the left-right axis; see Fig. 4.1 below) and hand ability. The other option is to forget about the strong-weak axis, the preference/ability distinction, and other aspects of laterality than handedness; and use only a left-right scale with a cut-off point for defining left-handedness (or two cut-off points if the ambidextrous category is included). The advantage of the first option is that one has a fairly rigorous concept of left-handedness that is true to the details of the phenomenon of handedness. The disadvantage is that it complicates the issue of handedness and makes the emphasis on left-handedness seem arbitrary, requiring a justification for carving left-handedness out of the big picture of human laterality. The advantage of the second option is that it is closer to the pre-scientific conception of left-handedness and does not so obviously beg for justification. The disadvantage is that it ignores things we do know about handedness and over-simplifies the phenomenon.

All scientific studies of left-handedness I have seen take the second option, avoiding the complexities of the strong-weak axis, the preference/ability distinction and the laterality of the rest of the body. Typically, handedness tests give a value on a graded scale (for example a whole number between 0 and 10, where 0 is totally left-handed and 10 is totally right-handed). A cut-off value is assigned for left-handedness (for instance, any individual score below 7 is considered left-handed and 7 or higher right-handed).

On the figure above, one would have only the left-right axis and anyone not falling in the "right-handed" category would be considered left-handed. Sometimes two cut-off values are assigned, giving three possible categories: left-handed, right-handed and ambidextrous. Whether one has the ambidextrous as a separate group, or groups everyone who is not strongly right-handed as left-handed, the determination of where the line is drawn between categories is fairly arbitrary. The standard against which handedness tests are measured is usually that when they are tested on a random population they should categorize close to 10% as left-handed. Tests are then calibrated by altering either questions or cut-off values (or both). This, in turn, makes claims to the effect that approximately 10% of humans (in the Western world at least) are left-handed rather circular.

Regardless of whether one looks at "the big picture" of laterality or sticks to simplified views of handedness, it should be clear that when it comes to classifying left-handers there are no seams at which to cut nature. There is a common-sense, or pre-scientific, conception of left-handedness, but when science tries to make this conception rigorous it blurs and fragments. It is therefore all the more surprising—and suspicious—that so many scientists ignore the complexity of laterality and insist on studying left-handers as if they form a particular kind, clearly delimited by nature.

The lack of a clear criterion for left-handedness has led some to look for the "essence" of left-handedness: some necessary and sufficient conditions that determine who is left-handed once and for all. One obvious place to look is the asymmetry of the brain. We know that the right hemisphere of the brain controls the movement of the left side of the body and the left hemisphere the movement of the right side. Since Paul Broca's discovery in the 1860s, we also know that our ability to speak resides in some significant way in the left hemisphere (McManus 2002, pp. 10–13). In other words, most of us are left-hemisphere dominant for speech. This may seem to be the solution then: Right-handers have a dominant left-hemisphere and left-handers, conversely, have a dominant right-hemisphere. Unfortunately, this solution is too simple. If one hemisphere is "put to sleep" by an injection of sodium amytal, a barbiturate tranquilizer, into the carotid artery on one side, half of the body becomes paralyzed (the side opposite to the hemisphere) and many mental functions stop or are slowed down. This test, called the sodium amytal test, can be used to determine which hemisphere controls speech functioning. This test has showed that only about 20% of left-handers and 3–4% of right-handers are right-hemisphere dominant for speech (Coren 1992, pp. 100–102; Corballis 1983, pp. 67–69). This has been confirmed by more recent studies using transcranial Doppler ultrasonography (McManus 2002, pp. 200–201). The overall laterality of the brain does therefore not strictly determine handedness. Right-hemisphere dominance is clearly much more common in left-handers than right-handers, but, nonetheless, 80% of left-handers are left-hemisphere dominant for speech.

There have been many speculations that left-handedness is caused by some neurological failure or damage to the brain. As right-handedness is the norm, it is assumed that when everything goes "right" in the neurological development of a child, it turns out right-handed. If something goes wrong, it may turn out left-handed. Many scientists, among them Stanley Coren, believe that some cases of left-handedness are *pathological* in this manner. Coren has tirelessly postulated that left-handedness is a marker for a variety of diseases and abnormalities, arguing that if something goes wrong in the neurological development so that a person turns out left-handed, then it is very likely that something else is wrong too (Coren 1992, pp. 153–164). He even coined the word "alinormal" to describe those double deviants who are "left-handed but otherwise normal."

Some scientists have tried to estimate the number of pathological left-handers; the results range from 5 to 50% of left-handers (Harris 1993b, p. 243). This, again, is all rather questionable, as very little is known about the causes of handedness. In general, it is surprising how little is actually known about left-handedness, despite enormous amounts of research. Even if the class of left-handers is to a great extent arbitrarily defined, that does not seem to explain this lack of results about the causes of left-handedness, or any other lateralization of the human body.

The scientific discourse on left-handedness arose in connection with a direct need of a power structure to increase the effectiveness of its control, and with negative associations of left-handedness always lurking in the background. Despite the absence of that need now, or at least its decrease and transformation, the discourse continues and it does have power effects. The power effects are not taking place in the class rooms and homes as would be expected if the power relations between educators and parents on the one hand and left-handed children on the other hand had not drastically changed. The power effects are less visible, less direct than tying up the left-hand of children, but they are still there. In particular, the scientific discourse on left-handedness often perpetuates the idea that left-handedness is bad, wrong or at least a sign of something gone wrong. These effects will become more visible, I hope, when we consider scientific studies on the longevity of left-handers and the resistance from left-handers to the power effects of this scientific research.

4.3 Left-Handed Science: The Longevity of Sinistrals

From 1988 to 1994, several attempts were made to find out whether left-handers die younger than right-handers. I am going to focus on three studies on the longevity of lefties. I shall look at attempts to produce and circulate as scientific knowledge the claim that left-handers die younger than right-handers, and I shall consider the power effects of this scientific endeavour and the resistance that took place. In short, I shall be describing (and inevitably be participating in) a politics of truth.

4.3.1 Why Think That Left-Handers May Have Reduced Longevity?

First of all, the question seems to be quite straightforward. Statistics about life expectancy are familiar to most people, showing, for instance, that females can expect to live longer than males. One would think it should be quite simple to determine whether there is any difference in life expectancy between left-handers and right-handers. Still, it is not obvious why this question should be posed in the first place. One can see how biological as well as social and cultural differences can account for the different life expectancy of males and females, but such differences between left-handers and right-handers are much less obvious. As left-handedness is defined pre-scientifically, that is, as a preference of using the left-hand for some salient tasks, there is nothing to indicate the possibility of a higher mortality risk for left-handers. Why would it occur to anyone that, say, if you prefer writing with your left hand you might expect to die younger than those who prefer using their right hand?

There are to two main reasons usually given for raising the question. First, some studies indicate that there are many more left-handers in younger age groups than the older ones. According to one study (Porac and Coren 1981), 13% of 20-year-olds are left handed, 5% of those in their fifties and virtually nobody of 80 or above. The other reason is that left-handedness has been associated with all sorts of health risks such as alcoholism, smoking, breast cancer and several neurological and immune disorders.[4] Left-handers may also be more prone to accidents (Coren 1989). According to Wayne P. London, left-handedness has been associated with other abnormalities, such as

> dyslexia, stuttering, hyperactivity or attention-deficit disorder (ADD), autism, schizophrenia, tardive dyskinesia, mental retardation, epilepsy, early-onset Alzheimer's disease, neural tube defects, skeletal abnormalities, severe prematurity, cleft palate, and chromosomal abnormalities. (London 1989, p. 1041)

And this list is still getting longer. It is striking how much of left-handedness research is concerned with various links between left-handedness and physical, mental, behavioural and social deviations. There has also been some, but much less, research conducted on the association of left-handedness with increased cognitive abilities and creativity, the "freakish" talents of left-handers which I discussed in the first section of this chapter.

Now let's go back to the pre-scientific definition: Left-handers prefer to use their left hand for writing and eating (or salient tasks of that sort). This sorts people quite clearly into left-handers and right-handers, clearly enough for everyday needs. But as I discussed in the previous section, when left-handers become the subject of study, it is not so clear anymore who is left-handed and who is not. In the studies which I shall look at, there is a strong tendency to define the subject so as to approach

---

[4]See in general London (1989), and in particular, association of left-handedness with alcoholism: London et al. (1985); with smoking: Harburg et al. (1978); with breast cancer: Titus-Ernstoff et al. (2000); with neurological disorder: Coren and Halpern (1991, p. 102); and with immune disorders: Geschwind and Behan (1982).

the pre-scientific human kind. This reveals yet again the concern with left-handers and left-handedness, as opposed to the more general concern with human laterality or cerebral asymmetry. In other words, scientists are not so much concerned with finding the truth about various kinds and degrees of laterality, but with finding the truth about left-handers.

It is sometimes claimed that the proportion of left-handers is reduced in older age-groups. Such claims often refer to findings reported in a 1981 book by Clare Porac and Stanley Coren. This observation can be explained in three ways:

(1) *The modification hypothesis*: There are fewer old left-handers because until recently many left-handers were forced to become right-handed. Younger left-handers are not confronted with the same pressure as the older ones were.
(2) *The elimination hypothesis*: There are fewer old left-handers because left-handers die younger than right-handers on average.
(3) *The bad sample hypothesis*: The numbers reported by Porac and Coren are wrong.

Stanley Coren has spent considerable effort conducting studies purporting to show that left-handers die younger than right-handers. Of the three studies I will be concerned with here, two were conducted by Stanley Coren and Diane Halpern. Both are psychologists, but Coren is best known for his popular books about dogs. The third study was conducted by Marcel E. Salive and Jack M. Guralnik who were with the Epidemiology, Demography and Biometry Program at the National Institute on Aging, National Institutes of Health in Maryland, and Robert J. Glynn who is at the Harvard Medical School.

4.3.2 Study I: The Baseball Encyclopedia (1988)

The first study by Halpern and Coren (1988) used a rather unorthodox source of data, namely *The Baseball Encyclopedia*. As Coren put it in an interview: "What we needed was a set of individuals, all of known handedness and all already dead. And that's not easy to find" (Waters 1989, p. 26). Coren and Halpern used records for 2271 major-league baseball players, excluding those baseball players whose handedness was ambiguous. The definition of left-handedness, therefore, is here that a player is classified as left-handed if he uses left hand for both throwing and batting throughout his career.

The mean age at death turned out to be approximately 8 months lower for left-handers than right-handers:

Results	Number	Mean age at death
Right-handed	1472	64.64 (s.d. = 15.5)
Left-handed	236	63.97 (s.d. = 15.4)

In 1988, Stanley Coren and Diane F. Halpern announced the results of their study in a letter to *Nature* (Halpern and Coren 1988). Although the apparent difference in mean age at death is not great, the results made the headlines of newspapers; radio and TV networks picked up the story and, of course, science magazines reported and discussed the findings. Articles on Coren's study appeared in as diverse publications as *Weight Watchers Magazine, Parents* (entitled "Attention, Parents of Left-Handers") and *Chemical and Engineering News* (Coren 1992, p. 213; Harris 1993a, p. 203; Reese 1988, p. 56). A new bit of scientific knowledge had been produced and was already in wide circulation.

In the scientific magazines, letters and articles poured in criticizing Coren and Halpern's study. Many letters pointed out that the data are not that representative of the normal population. One critic noted that "left-handed women cricketers need not worry unduly" (W.W. Nichols of John Radcliffe Hospital in Oxford, quoted in Reese 1988, p. 56). Other critics appealed to the modification hypothesis, that there are fewer left-handers to be found in older age groups simply because so many of them had been forced to become right-handed. Most of the criticism was concerned directly with Coren and Halpern's classification of handedness and their statistical reasoning, on which the whole study hinges. Max G. Anderson, described as "a left-handed number cruncher with the Canadian Statistical Analysis Service in Vancouver" applied different statistical methods and a slightly different way of classifying left-handed players, thereby including some ambidextrous players in the left-handed group. He used a more recent edition of a sports encyclopedia, and came to the opposite conclusion: left-handers live longer than right-handers (Weiss 1989, p. 180; see also Anderson 1989, p. 112).

Further criticism came from a study conducted by Elizabeth Wood at the California Institute of Technology. Like Anderson, she failed to replicate the results using a more recent edition of the *Baseball Encyclopedia*. She also evaluated Halpern and Coren's statistics and found that

> even if there were no difference in the mortality distributions of left-handers and right-handers, a difference in the cumulative survival fraction larger than that presented by Halpern and Coren would arise by chance in about 98 out of 100 samples of the size presented here. (Wood 1988, p. 212)

The conclusion drawn from her study was that in that limited sample "there is no statistically significant difference between mortalities of left-handed and right-handed baseball players" (Wood 1988, p. 212).

In his book *The Left-Hander Syndrome*, Stanley Coren makes some interesting remarks on this criticism. Throughout his book, whose target group appears to be non-scientist left-handers, he gives a rather idealistic picture of scientific practice: he is the rigorous scientist working in his laboratory and discovering things about his subject in accordance with strict scientific methods. Coren frequently talks about his "laboratory" and even "the experimental method." It is perhaps not very surprising that he gets somewhat annoyed when that self-image is broken and he gets criticism not from fellow scientists but from his subjects, the left-handers. Coren does not respond to their criticism, but questions their motives.

> Perhaps it was predictable that the two major comments on our findings were either authored or directed by left-handers and that none of the attacks was initiated by psychologists or neuropsychologists who knew the data that had led us to conduct the study in the first place. The initial comment came from Wood (1988), who is described in one article as being "neither left-handed nor a scientist" but whose analyses were directed by E. Sterl Phinney, a left-handed professor of theoretical astrophysics from the California Institute of Technology. Another broadside came from Anderson (1989), a left-handed free-lance statistician, working out of his home in Vancouver. Both comments pointed out some of the problems associated with working with *archival data*. (Coren 1992, p. 214)

Coren then goes on to explain what archival data is and the problems such data poses for this type of research. He notes that "when using archival data, the strength and reliability of results that the researcher gets depend [...] upon the definition of handedness and the specific statistical tests used to determine the significance of the findings" (Coren 1992, p. 214). Coren and Halpern had already made a comment to this effect in a lengthy review published in the *Psychological Bulletin* in 1991, only a few months before his book was published. Referring to the articles by Anderson (1989) and Wood (1988), they write: "It appears that the statistical significance of these results varies as function of the specific statistical analyses used and as a function of the definition of handedness employed" (Coren and Halpern 1991, p. 214). What he does not say is that all attempts to replicate his results from the same or similar sources (in particular sports encyclopedias) had failed. The statistical methods he chose and the definition he employed happened to be the only ones to show that there is any considerable difference in the longevity of lefties and righties at all.

As Coren grudgingly pointed out, much of the criticism he faced came directly or indirectly from left-handers who were not psychologists, but well-versed in statistical methods. His scientific authority was challenged by his research subjects, leading to a struggle that took place outside his scientific discipline but still on scientific grounds. Coren's reaction to the criticism, as could be seen in the quote from his book above, was to reassert his scientific authority and dismiss his critics with ad hominem attacks: they are outsiders from his discipline; they are left-handers and hence biased; one critic is not a scientist and another critic is working out of his home.

4.3.3 Study II: The California Study (1991)

Stanley Coren and Diane Halpern decided to conduct another study, based on a more reliable set of data. They obtained death certificates from two counties in southern California (Coren and Halpern 1991). Two thousand questionnaires concerning the handedness of the deceased family member were sent to the listed next of kin, which resulted in 987 usable cases. They asked which hand the deceased used for writing, drawing and throwing a ball. If the deceased used the right hand for all the tasks, he or she was classified as right-handed, otherwise as left-handed. Coren notes that this classification includes ambidextrous in the left-handed group. He insists on calling the group left-handed, although "non-right-handed" would be a more appropriate label. Coren surprisingly remarks that he and Halpern "were insisting that the person be strongly right-handed to be counted as a right-hander, much as we did with the baseball player data" (Coren 1992, p. 218). On the contrary, the baseball player study classified left-handers as *strongly* left-handed, not as left-handed and ambidextrous.

In 1991, three years after the baseball study, Coren and Halpern published the results of their Californian study in *The New England Journal of Medicine* (Coren and Halpern 1991). Their result was that left-handers die on average 9 years earlier than right-handers:

Mean age of death	Men	Women
Right-handers	72 years 4 months	77 years 8 months
Left-handers	62 years 3 months	72 years 10 months

This second study of Coren and Halpern did not get any warmer reception than the first one. It was widely criticized and even ridiculed. Again the statistical methods were found faulty. Tricia Hartge, an epidemiologist at the National Institutes of Health, remarked that the results "mean absolutely nothing" because Coren and Halpern never found out the number of the left- and right-handed in those two counties, nor their age distribution. Actuaries ridiculed the study. The American Academy of Actuaries issued a statement calling the statistical methods faulty and Richard Labombarde, an actuary with the private firm of Milliman and Robertson in Washington, is quoted in the *New Scientist* saying that the research reminded him of methods used by actuaries 200 or 300 years ago (Charles 1991, p. 21).

The strongest criticism comes from the epidemiologist who conducted the third study I will discuss here, Marcel E. Salive. He noted that if the age distribution was anywhere close to what Porac and Coren's (1981) study had indicated, that is, if left-handers are on average so much younger than right-handers, and if the death rate is *the same* for the two groups, the difference in mean age at death should be even greater than what Coren and Halpern found in their California study. In other words, the age difference may be the cause and not the result of difference in mean age at death. Therefore, the difference in mean age at death does not by itself indicate what the difference in life expectancy may be, or whether there is a difference in life expectancy (Salive et al. 1993).

Chris McManus directs quite the same criticism at the study in his *Right Hand, Left Hand*, discussing the study in a chapter called "Vulgar Errors" (McManus 2002, pp. 292–293). Noting that left-handers are on average younger than right-handers (since there are more of them now than a few decades ago), he points out:

> People who read Harry Potter books are younger than those who do not. Ask the relatives of a group of recently deceased people whether their loved one had read Harry Potter and inevitably one will find a younger age at death in the Harry Potter enthusiasts, but that is

only because Harry Potter readers are younger overall. There is no need for a government health warning on the cover of Harry Potter books. (McManus 2002, p. 293)

Despite all this criticism, the findings were no less widely publicized than the earlier findings. Again the scientific knowledge had been produced and put into circulation, whatever one may think about its reliability or of the way it was produced.

I do not know to what extent this criticism came from left-handers. Resistance from left-handers did, however, take a non-scientific form when some left-handers started phoning in death threats (Holden 1991, p. 916). This reaction, this apparent anger of left-handers, may not be a very strong move strategically, but it does show that this production of scientific knowledge was having a considerable effect on the subjects of the study. We can call these effects power effects, for they are obviously affecting the behaviour and self-image of left-handers. The philosophical point to note here is that the conflict is not about the application of science, but about certain scientific knowledge and its immediate effects. The resistance to this knowledge production is not concerned with a risky application or misuse of scientific results. It is concerned with the knowledge production itself. I shall come back to this resistance below.

After this second study and the criticism it got, science journals such as *Nature* and *New Scientist* observed that in order to produce the knowledge in question, other methods are needed (Pool 1991, p. 545; Charles 1991, p. 21). It is necessary to conduct a cohort study, that is, to take a group of right-handers and a group of left-handers and keep track of how many die from each group over a considerable length of time.

4.3.4 Study III: Marcel E. Salive et al. (1993)

The third study I will look at was conducted by epidemiologists Marcel E. Salive, Jack M. Guralnik and Robert J. Glynn. They analysed data from 3809 people aged 65 and older and recorded the mortality rate over six years. Subjects were classified "as right-handed if they reported using their right hand for both writing and cutting with scissors and as left-handed if they used their left hand or either hand for either task" (Salive et al. 1993, p. 265). Salive classifies the subjects as left-handed if they are not strongly right-handed. As he remarks himself, this classification principle is comparable to that used in Coren and Halpern's (1991) California study and in Porac and Coren's (1981) study. It is, however, different from the classification used in Halpern's and Coren's *Baseball Encyclopedia* study (1988), where a player is classified as left-handed if he uses left hand for both throwing and batting throughout his career.

Salive found left-handedness not related to age, contrary to Porac and Coren (1981)—the study which initiated this line of research:

The six year mortality rate turned out to be almost the same for the right-handed and the non-right-handed: Over the six years (1982–1988) 33.8% of the left-handers

Results	By age group			By gender	
	65–74	75–84	85+	Men	Women
Ratio of left-handers	6.9%	7.3%	7.3%	9.1%	5.8%

died and 32.2% of the right-handers. The difference is not statistically significant. These results, therefore, support the null hypothesis that there is no difference in the mortality rate of right-handers and left-handers.

Coren and Halpern's reaction to this study was to claim:

> It is a basic principle of the experimental method that failures to reject the null hypothesis cannot be used to cancel statistically significant results. There are many ways to obtain results that are not statistically significant, including sloppy research and low statistical power. (Coren and Halpern 1993, p. 238)

Salive's study, they say, falls under the latter, namely low statistical power. It is worth noting, however, that Salive's sample was almost four times as big as Coren and Halpern's sample in their California study, and nearly 70% larger than the sample in their baseball study. Apart from Coren and Halpern's criticism, the reactions to Salive's study were minimal and it received no publicity in the popular media.

4.4 Left-Handers and Knowledge Production

In studies on left-handers the definition of handedness varies and so it is not very clear what the subject of study is exactly. In the various baseball studies, the left-handed are usually taken to be those who both bat and throw with their left hand, although some of the "replications" included the ambidextrous to some extent. The other two studies take "left-handed" as "non-right-handed," so that *any* deviation from consistent right-handedness is classified as left-handedness. The definition of handedness naturally affects studies on left-handers. A study from 1989 conducted by Stanley Coren, indicated that left-handers were more prone to accidents than right-handers (Coren 1989, p. 1040). He used a four-item self-report inventory to classify subjects by handedness. The items were: drawing, throwing a ball, dealing cards and using an eraser on paper. Coren does not say how exactly handedness was determined from this inventory, but the result is a two group classification and the left-handed class reports having accidents more frequently than the right handed one (Coren 1989, p. 1040).

Another study regarding handedness and accidents, from 1993, uses a graded scale for handedness, but still distinguishing three groups: left-handed, right-handed and ambidextrous. This study indicates that the *ambidextrous* are more prone to accidents, but left-handers do not have more accidents than right-handers (Hicks et al. 1993). Other studies have found stronger associations with "mixed handedness" or weak handedness, than with left-handedness. In those cases left-handers and right-handers turn out to have no statistical difference for the association (pathology, abnormality etc.) in question, while the ambidextrous show a stronger association.

The difficulty scientists have had producing knowledge about left-handers is evident from the studies on the longevity of left-handers. This difficulty is also borne out by other research on left-handers, where conflicting results are the rule rather than the exception. But conflicting studies on left-handers keep piling up, regardless of the difficulties and the apparent lack of scientific rationale for researching left-handedness specifically and outside the context of human laterality.

I find it suspect, keeping in mind the history of left-handedness, that researchers single out left-handers as a group ignoring to various degree the complexity of handedness—and indeed the complexities of human laterality. Given the known complexities, which I described in the previous section, and the research mentioned above that finds an association with "mixed handedness" but not with left-handedness, not to mention the failure to produce reliable knowledge about left-handers, lefties studies must be regarded as highly suspect. This point is, of course, a "bad science" type of criticism: the object of the scientific study, left-handedness or left-handers, is poorly defined and an improper object for scientific study. But I am taking this point further: If a researcher would rigorously define "left-handedness," taking into account all the known complexities of human laterality, he would be producing better scientific knowledge, but there would still be a reason for left-handers to resist this knowledge production. Not because it is bad science, nor because of risky applications or misuse of the results, but because it classifies a certain group as left-handers according to criteria that are to an important extent arbitrary, and then represents this group as biological failures.

Despite the fact that the only evidence for the claim that left-handers die younger than right-handers are the two studies by Coren and Halpern, and despite the criticism, the presence of numerous studies failing to replicate the results and the absence of studies replicating the results; it is still widely believed that left-handers do on average die younger than right-handers. This has to do with how scientific knowledge is produced, circulated, diffused and consumed; that is, the features of the modern régime of truth. In my summary of the three studies on the longevity of lefties, I discussed the production of this knowledge and its circulation through scientific journals. I only briefly touched upon how this knowledge is diffused and consumed by the public. In the case of the longevity research, the diffusion and consumption is quite interesting. The results of the two positive studies were presented not only in scientific journals, but also in all sorts of popular and semi-popular magazines (like *Weight Watchers Magazine*, *Parents* and *Chemical and Engineering News*). The results were also discussed in the mass media, culminating perhaps in Stanley Coren's appearance on an American television talk show (including the mandatory confrontation with a proud but upset left-hander and a cheering/booing studio audience). The mass media, and also the magazines that are between mass media and academic journals, do not report scientific results unless it makes a good story. Positive findings are a much better story than negative findings, just as war makes a much better story than peace. For this reason, positive results about associations with left-handedness are frequently reported by the mass media, while negative results do not get reported. The story about the reduced longevity of left-handers was a good story not only because it was a story about positive scientific findings, but also because it manifests the interest that left-handers and left-handedness still arouses among the public—which again has its historical reasons as I discussed in the first section of this chapter.

As an example of the "good story bias" of the mass media, a lengthy article on left-handers in the December 1994 issue of the *Smithsonian* portrays Stanley Coren as an authority on scientific studies of left-handers and their longevity (Shute 1994). The article introduces Salive's study only to have Coren reject it by citing more if his own studies. Lauren Julius Harris, who has criticized Coren's studies more thoroughly than anyone else, is not mentioned at all. A stronger example of the "good story bias" is of course all the media hype at the time when Coren and Halpern published their two longevity studies. Salive's study, as I noted earlier, went largely unnoticed by the mass media. This is an important feature of how scientific knowledge is currently circulated and consumed in our societies.

In Sect. 3.4 I discussed the five features of the modern régime of truth, according to Foucault's analysis. I have mentioned some of these points in this chapter in relation to the scientific study of left-handers, but let me now go through them point by point.

First point: The truth about left-handers *is centred on the form of scientific discourse and the institutions which produce it*. The supreme source of truth about left-handedness and left-handers is science (primarily psychology) and its institutions. I mentioned one example above, that of the New Jersey superintendent who claimed not to have enough knowledge about left-handers to know how to deal with them. Another example would be the fact that even left-handers are expecting science to tell them who is *really* left-handed. It is not enough to consider oneself left-handed as a result of being recognized as one in everyday situations. There "must be" a scientific basis for this kind of humans, an essence of left-handedness, and once that is discovered we shall know who the real left-handers are and who just appear to be left-handed. In the light of my discussion of the fate of left-handedness in scientific studies, I question this reliance on scientific authority to tell us anything about left-handedness.

Second point: Scientific knowledge about left-handers *is subject to constant economic and political incitement*. The political incitement, in particular the need to manage, control and discipline left-handed children at school and at home, has decreased, while the economic incitement, in particular to deal with left-handers as consumers and as children with "special needs," has increased. The economic need to deal with left-handers is not just a matter of producing tools or other objects that are adjusted to the needs of left-handers. It is also a need for parents and teachers to deal with left-handed children—not to discipline them but to manage them—and this need is in a context that is increasingly commercial. On one hand, the educational systems in Western countries are becoming increasingly commercialized, and, on the other hand, there is a growing industry that caters to the needs of parents, not only with toys, clothes, and various products for children; but with books, magazines, courses and all sorts of services for parents. Parents of left-handed children are an important economic niche, the exploitation of which leads to considerable incitement for knowledge about left-handers.

Third point: Scientific knowledge about left-handers *is the object, under diverse forms, of immense diffusion and consumption.* I mentioned above the very diverse publications in which results of the first two studies appeared. The two studies by Halpern and Coren appeared in *Nature* and *The New England Journal of Medicine*, both highly regarded science journals. Criticism and discussion of the findings appeared in scientific journals such as *Science, Nature,* and, in particular, the psychology journals *Psychological Bulletin* and *Perceptual Motor Skills*; but also in science magazines like *New Scientist, Scientific American,* and *Smithsonian.* Reports of the results appeared in diverse popular magazines, newspapers and on television. I discussed above one interesting feature of modern diffusion and consumption of scientific knowledge, namely what I am calling "the good story bias."

Fourth point: Scientific knowledge about left-handers *is produced and transmitted under the control, dominant if not exclusive, of a few great political and economic apparatuses.* Foucault mentioned four such apparatuses: university, army, writing and media; and I wanted to add the fifth: corporations. In the case of left-handers science the relevant apparatuses are at least the university, writing and media. The media aspect overlaps with the previous point regarding the diffusion and consumption of knowledge, and the university aspect overlaps with the first point about truth being centred on scientific discourse and the university. The emphasis of the first point, however, is that the supreme source of truth is science and its institutions, while the emphasis of this fourth point is on the control of the distribution of knowledge. Knowledge about lefties comes from scientific studies, which are conducted in the conventional university environment. One illustration of this control is the reaction of Stanley Coren when he got criticism from left-handers, non-psychologists, a non-scientist and even a statistician "working out of his home." His were ad hominem attacks, but it is revealing that Coren objects exactly to the point that the criticism is not originating in the proper institutions (which would be university psychology departments).

Fifth point: Scientific knowledge about left-handers is *the issue of a whole political debate and social confrontation.* The "political debate" is here political in the sense of Foucault's power relations. There is a resistance from left-handers to certain knowledge production about them, resulting in a conflict with the researchers. This resistance and conflict has primarily taken the form of scientific criticism (attacking studies for being bad science) and death threats. I have tried in this chapter to provide a different kind of resistance. This resistance is based on awareness of the history of left-handers science and of the details of the knowledge production itself, as well as an analysis of the present "politics of truth" (that is, the dynamics between researchers of left-handers and the left-handed subjects). The motivation of this critique is not merely to challenge poor scientific practice, or bad science, but to resist the effects of this science, which represents left-handers as not only abnormal, but as a sort of biological failures: sick, accident prone, and short-lived.

4.5 Conclusion

Left-handedness has throughout recorded history caused curiosity and even repulsion, and it has been fought by parents and educators—probably since parents and educators emerged as people with their own specific roles in Western cultures and societies. The view of left-handedness and left-handers got a new dimension with the rise of statistics and statistical reasoning: left-handedness became bad in a new way, it became *abnormal*. This change also marks the beginning of the career of left-handedness as a scientific concept.

The concept of left-handedness has a long and widespread history of negative associations and, I claim, this has affected recent scientific research on left-handedness in at least two ways: left-handedness became something that needed to be studied and was interesting to study as a deviation, and there has been a strong tendency to consider left-handedness as a symptom of deficiency or developmental failures. There is also a need to deal with left-handers in the schools and at home, which calls for increased research on left-handedness—manifesting a dynamics of knowledge and power. The need to deal with left-handers is still there, although now it is not so much concerned with changing them (although it probably would be if acceptable methods were available). Parents and educators, and now also employers and even designers and marketers, still need to know what to do about left-handed people.

I have tried to show how the history of the concept of left-handedness has affected the scientific discourse on left-handedness and the troubles of the concept of left-handedness in scientific studies. I have also sketched the relations of power that gave rise to scientific studies on left-handers and the relations of power that appeared while scientists tried to find an answer to the question of whether left-handers die on average younger than right-handers. Furthermore, I have tried to free us from thinking of left-handedness as something obvious, as a clear and necessary category given by nature.

My discussion in this chapter is not only an analysis of a politics of truth, but also a form of critique. I question power on its discourse of truth: (1) I point to the history of left-handedness and question the motives of those who want to study left-handers—in particular with regard to longevity. (2) I criticize the epistemology of left-handers studies, by arguing that the concept of left-handedness is inappropriate for scientific study. (3) I question the need to ask science for the truth about left-handers, rather than simply relying on personal experience (for example, when asking who is left-handed). I also question the discourse of truth on its power effects: Given that left-handedness is to an important extent arbitrarily defined, and that the scientific study of left-handers has failed miserably in producing reliable and reproducible results, the scientific discourse on left-handers should not be having the sort of power effects that it does, it should not be convincing anyone that left-handers die earlier than right-handers, or that left-handers are in any sense biological failures.

References

Anderson, Max G. 1989. Lateral Preference and Longevity. *Nature* 341: 112.

Árnason, Garðar. 2017. Biopolitics and the Longevity of Left-Handers. In *Bioethics and Biopolitics*, ed. Péter Kakuk, 59–76. Dordrecht: Springer. https://doi.org/10.1007/978-3-319-66249-7_5.

Binks, Georgie. 2008. Why Are So Many U.S. Presidents Southpaws? *CBCNews* 7 March. http://www.cbc.ca/news2/background/science/left-handed-presidents.html. Accessed 13 Sept 2018.

Bishop, D.V.M. 1989. Does Hand Proficiency Determine Hand Preference? *British Journal of Psychology* 80: 191–199.

Blau, Abram. 1946. *The Master Hand: A Study of the Origin and Meaning of Right and Left Sidedness and its Relation to Personality and Language*. New York: The American Orthopsychiatric Association Inc.

Brown, Susan G., Eric A. Roy, Linda E. Rohr, Benjamin R. Snider, and Pamela J. Bryden. 2004. Preference and Performance Measures of Handedness. *Brain and Cognition* 55: 283–285.

Burt, Cyril. 1937. *The Backward Child*. New York: Appleton.

Charles, Dan. 1991. Left-Handers Don't Die Young After All. *New Scientist* 1766: 21.

Cholmeley, Roger James (ed.). 1919. *The Idylls of Theocritus*. London: G. Bell & Sons.

Corballis, Michael C. 1983. *Human Laterality*. New York: Academic Press.

Corballis, Michael C. 1991. *The Lopsided Ape*. Oxford and New York: Oxford University Press.

Coren, Stanley. 1989. Left-Handedness and Accident-Related Injury Risk. *American Journal of Public Health* 79: 1040–1041.

Coren, Stanley. 1992. *The Left-Hander Syndrome: The Causes and Consequences of Left-Handedness*. New York: The Free Press.

Coren, Stanley, and Diane Halpern. 1991. Left-Handedness: A Marker for Decreased Survival Fitness. *Psychological Bulletin* 109: 90–106.

Coren, Stanley, and Diane Halpern. 1993. Left-Handedness and Life Span: A Reply to Harris. *Psychological Bulletin* 114: 235–241.

Corey, David M., Megan M. Hurley, and Anne L. Foundas. 2001. Right and Left Handedness Defined: A Multivariate Approach Using Hand Preference and Hand Performance Measures. *Neuropsychiatry, Neuropsychology, and Behavioral Neurology* 14 (3): 144–152.

Dawson, Binnie, and L.M. John. 1977. An Anthropological Perspective on the Evolution and Lateralization of the Brain. *Annals of the New York Academy of Sciences* 299: 424–447.

Frayer, David W., Marina Lozano, M. José, Bermúdez de Castro, Eudald Carbonell, Juan Luis Arsuaga, Jakov Radovčić, Ivana Fiore, and Luca Bondioli. 2012. More Than 500,000 Years of Right-Handedness in Europe. *Laterality: Asymmetries of Body Brain and Cognition* 17: 51–69.

Gay, John. 2005 [1727]. *Fables: With a Memoir by Austin Dobson*. Whitefish, MT: Kessinger Publishing.

Geschwind, Norman, and Peter Behan. 1982. Left-Handedness: Association with Immune Disease, Migraine and Developmental Learning Disorder. *Proceedings of the National Academy of Sciences* 79: 5097–5100.

Göbel, Wolfgang, and Michaela Glöckler. 2015. *Kindersprechstunde: ein medizinisch-pädagogischer Ratgeber*, 19th ed. Stuttgart: Urachhaus.

Halpern, Diane, and Stanley Coren. 1988. Do Right-Handers Live Longer? *Nature* 333: 213.

Harburg, Ernest, Anna Feldstein, and James Papsdorf. 1978. Handedness and Smoking. *Perceptual and Motor Skills* 47: 1171–1174.

Harris, Lauren Julius. 1980. Left-Handedness: Early Theories, Facts and Fancies. In *Neuropsychology of Left-Handedness*, ed. Jeannine Herron, 3–78. New York: Academic Press.

Harris, Lauren Julius. 1993a. Do Left-Handers Die Sooner than Right-Handers? Commentary on Coren and Halpern's (1991) 'Left-Handedness: A Marker for Decreased Survival Fitness'. *Psychological Bulletin* 114: 203–234.

Harris, Lauren Julius. 1993b. Reply to Halpern and Coren. *Psychological Bulletin* 114: 242–247.

Hertz, Robert. 2013 [1909]. The pre-eminence of the right hand. A study in religious polarity. *HAU: Journal of Ethnographic Theory* 3 (2), pp. 335–357. The essay was first published in 1909 as

"La prééminence de la main droite: étude sur la polarité religieuse," *Revue Philosophique* 68: 553–580.

Hicks, Robert A., Karen Pass, Hope Freeman, Jose Bautista, and Crystal Johnson. 1993. Handedness and Accidents with Injury. *Perceptual and Motor Skills* 77: 1119–1122.

Holden, Constance. 1991. Slugging it Out over Left-Handed Mortality. *Science* 252: 916.

Kierkegaard, Søren. 1965 [1841]. *The Concept of Irony*. Bloomington, IN: Indiana University Press. Trans. L.M. Capel.

Liddell, Henry George, and Robert Scott. 1992. *Intermediate Greek-English Lexicon*, 7th ed. Oxford: Oxford University Press.

Lloyd, Geoffrey Ernest Richard. 1962. Right and Left in Greek Philosophy. *Journal of Hellenic Studies* 82: 56–66.

London, Wayne P. 1989. Left-Handedness and Life Expectancy. *Perceptual and Motor Skills* 68: 1040–1042.

London, Wayne P., Priscilla Kibbee, and Laurent Holt. 1985. Handedness and Alcoholism. *The Journal of Nervous and Mental Disease* 173: 570–572.

McManus, Chris. 2002. *Right Hand, Left Hand: The Origins of Asymmetry in Brains, Bodies, Atoms and Cultures*. Cambridge, MA: Harvard University Press.

McMullin, Walter G. 1914. What Shall We Do with Left-handed Pupils? A Symposium Conducted by Walter G. McMullin. *Teacher* 18: 331–338.

Needham, Rodney. 1967. Right and Left in Nyoro Symbolic Classification. *Africa* 37: 425–451.

Parma, Valentina, Romain Brasselet, Stefania Zoia, Maria Bulgheroni, and Umberto Castiello. 2017. The Origin of Human Handedness and its Role in Pre-Birth Motor Control. *Scientific Reports* 7: 16804. https://doi.org/10.1038/s41598-017-16827-y.

Pool, Robert. 1991. Can Lefties Study be Right? *Nature* 350: 545.

Porac, Clare, and Stanley Coren. 1981. *Lateral Preferences and Human Behaviour*. New York: Springer-Verlag.

Reese, Kenneth M. 1988. Do Right-Handers Live Longer than Left-Handers? *Chemical & Engineering News* 66 (36): 56.

Rigby, Peter. 1966. Dual symbolic classification among the Gogo of Central Tanzania. *Africa* 36: 1–16.

Salive, Marcel E., Jack M. Guralnik, and Robert J. Glynn. 1993. Left-Handedness and Mortality. *American Journal of Public Health* 83: 265–267.

Shute, Nancy. 1994. Life for Lefties: from Annoying to Downright Risky. *Smithsonian* December: 131–143.

Titus-Ernstoff, Linda, et al. 2000. Left-handedness in relation to breast cancer risk in postmenopausal women. *Epidemiology* 11: 181–184.

Uomini, Natalie T. 2009. The Prehistory of Handedness: Archaeological Data and Comparative Ethology. *Journal of Human Evolution* 57: 411–419.

Waters, T. 1989. Sinistral Stats. *Discover* (April): 26.

Weiss, Rick. 1989. Lefties and Longevity: Look Again. *Science News* 136 (12): 180.

Wilbur, James Benjamin, and Harold Joseph Allen. 1979. *The Worlds of the Early Greek Philosophers*. New York: Promotheus Books.

Wilson, Frank R. 1998. *The Hand: How Its Use Shapes the Brain, Language, and Human Culture*. New York: Pantheon Books.

Woo(Ding-Liang), T.L., and Karl Pearson. 1927. Dextrality and Sinistrality of Hand and Eye. *Biometrika* 19: 165–199.

Wood, Elizabeth K. 1988. Less Sinister Statistics from Baseball Records. *Nature* 335: 212.

Chapter 5
Icelanders as Subjects of Science

In this chapter I discuss plans that were made in Iceland to establish a database with health, genetic and genealogical data for the entire nation. I first describe these plans in some detail and then discuss their relation to the discredited eugenics discourse and how that discourse together with the promises of a genetics revolution was deployed to produce a docile research population. In this case the Foucauldian biopolitics of the population is concentrated on a population of a single nation as a research population, for the benefit of science and the economy.

5.1 Introduction

Left-handers form a group, which at first glance appears to lend itself to scientific study.[1] In the previous chapter I argued, however, that when one tries to become clear about the concept of left-handedness as a scientific concept, it becomes very problematic. The point was, though, not merely that science on left-handers is somehow wrong because its object of study is poorly defined or even in some sense made up, the point was more importantly that science on left-handers is worth opposing, at least by left-handers, because of the specific relations of power and knowledge at work in this case.

Icelanders do not seem to form a group which lends itself to scientific study, simply because nationality is obviously a property that has no biological basis. Still, Icelanders have been considered a very interesting subject of study. One project, which will be the focus of this chapter, planned to collect health data, genetic data and genealogical data from the entire population in order to conduct research in

---

[1]Parts of this chapter, in particular from Sects. 5.3 and 5.4, have appeared in the book chapter "Interbreeding within the Icelandic population is high compared to that of mice or fruit-flies" (Árnason 2004). I thank the University of Iceland Press/Háskólaútgáfan for the permission to reprint material from that chapter.

© The Author(s) 2018
G. Árnason, *Foucault and the Human Subject of Science*, SpringerBriefs in Ethics,
https://doi.org/10.1007/978-3-030-02813-8_5

genetics. At the end of the 20th century, there was a common feeling in Iceland and elsewhere that a new age was dawning, the age of genetics, which would bring cures to some of the most common diseases causing death in developed countries. Those who opposed the project on the grounds that it conflicted with conventional research ethics were dismissed as being behind the times. For instance, Icelandic Director General of Public Health, Sigurður Guðmundsson, said in an interview:

> We are walking into a new world. But I don't think it is wild to say this may turn out to be a tool like none other. And I don't think this country can just sit here and say, Nope, sorry, we are going to stand on rules that existed in a different era for a different world. (Specter 1999, p. 51)

This view, that genetics was leading us into a new world at the dawn of a new century, was not limited to Iceland. Some claimed that a revolution was taking place, the Genetic Revolution, comparable to the Industrial Revolution in its impact on our societies (e.g., Russo and Cove 1998, p. ix; also Greely 1992, and Greely 2001). Genetic science and technology have indeed been growing at a spectacular rate since the 1990s. And so, for the last twenty years, genetics has been on the verge of transforming medicine. The record so far, however, is not very encouraging.

In this chapter I will focus on plans that were floated in Iceland in the late 1990s to create a single health sector database for the entire nation, intended in part for research in human genomics and genetic epidemiology. The Health Sector Database (HSD) was to contain information gathered from medical records and be connected to, or merged with, two other databases: one containing genealogical data for every Icelander alive and, going centuries back, most of those deceased; and the other containing genetic information. A private company, deCODE genetics, was given exclusive rights for 12 years to construct the database, maintain it and sell access to the information. The company is usually referred to as "deCODE genetics" in English texts, but "Íslensk erfðagreining" (literally "Icelandic Genetic Analysis") in Icelandic texts. deCODE genetics was incorporated in Delaware, U.S.A., and later bought by biopharmaceutical giant Amgen, but it is located in Reykjavík, Iceland, and is still operating.

On December 3–7, 1999, The American Society of Hematology held its 41st annual meeting and exposition in New Orleans. A plenary policy forum brought together Kári Stefánsson, CEO of deCODE genetics, prominent bioethicist and professor of health law George Annas, and the director of the US National Institutes of Health Francis S. Collins. Much of the enthusiasm, and the thought behind the project, is captured in the following quote from an announcement of the plenary policy forum:

> How individuals are born with a pre-determined color of eyes and hair and with or without freckles are examples of a basic level of genetics that most of us study in introductory biology. Medical researchers who focus on genetics, however, sift daily through genetic code searching for explanation of diseases including Alzheimer's, heart disease and cancer. This ongoing study has provided the foundation for the new discipline of genomics and gene therapy. Dr. Kári Stefánsson has taken a giant leap forward in this field. He is the founder of Iceland's first biotechnology company, deCODE Genetics. Dr. Stefánsson has the support of Iceland's government for deCODE to run an unparalleled database that will collect, store,

package and sell the genetic heritage of Iceland, a nation of the most genetically homogenous people on earth. (The American Society of Hematology 1999)

On December 17, 1998, the Icelandic parliament passed a controversial bill, making the changes to Icelandic laws required for the health sector database project to go ahead. During the following weeks the international media reported that the Icelandic nation had "sold its genes." A headline in *The Star Tribune* read "Iceland: The Selling of a Nation's Genetic Code" (Crosby 1999), a CNN bioethics column on the Internet was entitled "Attention Shoppers: Special Today—Iceland's DNA" (Kahn 1999), *The Washington Post* announced "Iceland to Make its Genetic Code a Commodity" (Schwartz 1999a), on the cover of the January 18, 1999, issue of *The New Yorker* a headline reads "Genes For Sale" (Specter 1999), and the list goes on.

While the international media were mostly concerned with the commercialization of Iceland's genetic information, the debate in Iceland was more concerned with four ethical issues. I will not be particularly concerned with these ethical issues here, but it is worth noting what they are if only to serve as a contrast to the issues I will discuss. The first issue and perhaps the one that got most attention in Iceland is privacy. Information from medical records is confidential and it was from the beginning debated whether privacy and confidentiality could be guaranteed: deCODE boasted they would have the best coding system of its kind; critics object that the best is not good enough.

The second issue is consent. The principle for gathering medical information is that of "presumed consent." If you do not send in a specific form asking not to be included in the health sector database, your consent is presumed and information will be collected from your medical records. However, collecting genetic data for scientific research will require informed consent.

The third issue is that of freedom of research: deCODE was given exclusive rights to the medical information to be collected in the database. No one would be able use that information for scientific research without permission from deCODE, and it is explicitly stated that a permission will be given only if the proposed research is not perceived to conflict with deCODE's business interests.

The last issue is genetic discrimination. Could health insurance companies use genetic information, gained directly or indirectly from the database, to determine rates or deny a person insurance? At this time the Icelandic health care system is for the most part public, offering universal health care with moderate user fees, but the trend is towards privatization.[2]

These are fairly typical bioethical issues. The issues of privacy and consent of research subjects and the freedom of scientific research concern research ethics and proper scientific procedures. The last issue is an example of a questionable application of science.

Instead of a narrowly "ethical" focus on these kinds of issues, I propose in the following sections, as I did in the previous chapter, to engage in a critique of the type, which I am calling "politics of truth." The general structure of the critique is

---

[2]For a discussion of these issues, in particular the first three, see McInnis (1999), Chadwick (1999), and Häyry et al. (2007).

in this case similar to the left-handers' case: I identify and discuss certain aspects of a dynamic between a scientific discourse and a system of power (although it may be not so much a "system" of power as simply a diverse set of power relations and power effects), in order to resist some of these power effects and change the dynamics of some of the power relations. I analyse, criticize and question the scientific discourses at play and the power effects they have both in the plans to establish the Health Sector Database in Iceland and the representation of the scientific subjects, the Icelanders. My critical analysis focuses on the plans to establish the Health Sector Database with an emphasis on certain historical parallels, the effects the discourse around the HSD has had on the representation of Icelanders as subjects of science and the resistance in Iceland to this project.

In the following section I describe the Health Sector Database project in some detail. The third section discusses the roots of modern genomics in eugenics and similarities in the scientific discourse of eugenics and the scientific discourse of modern genomics in the context of the Icelandic database project. The fourth section considers how the techno-scientific database discourse turns the Icelandic population, quite literally, into a totally informative population of laboratory animals. The aim of this chapter is, again, to analyse and participate in a politics of truth. To paraphrase Foucault: I question the scientific discourse (human genetics) on its power and I question power on its discourse of science. Awareness of the power relations at work is already the first step towards resisting their subjugation. My hope is that one will see, as a result of my discussion, the issue in a new way, with the modest goal of better defending oneself and, perhaps less modestly, to change power relations between the scientific discourse of human genetics and the Icelandic lab animals, in favour of the latter.

5.2   The Icelandic Health Sector Database: The Plan

The Icelandic Health Sector Database was to contain information taken from medical records in Iceland. According to deCODE's plans, regular staff at medical institutions transfer data from medical records to medical information software designed especially for the purpose of processing the data and transferring it to the database. It is not clear what information is included, but, according to deCODE, it is most medically relevant data that can be coded, such as main symptoms, test results, diagnosis, stage and duration of disease, medication, treatment, results, side-effects and cost of treatment, who provides treatment, where and for how long, and some personal information, such as the patient's age group. Patients' narratives is not included, nor personal information that is not medically relevant. It is not clear whether information about diet, life-style (other than smoking and drinking habits), sexual orientation, race, vocation or income bracket is included, all of which can be medically relevant. deCODE's information booklet, "Health Sector Database: Questions and Answers," lists some sorts of information which will be entered into the database and then adds:

"Other factors in the patient's history that may be significant for his/her health, and can be coded, would also be entered in the database" (Íslensk erfðagreining 1998, p. 9).

The database laws, as they were approved by the parliament, make clear that the Health Sector Database may only include data from medical records, but it could be linked with other databases containing genealogical and genetic data (with the approval of the Data Protection Commission). Judging from deCODE literature (Íslensk erfðagreining 1998, p. 13), it seems that there were to be not three databases but a single database with three divisions. The difference may seem trivial, but it is important for two reasons. First, proponents of the Database project frequently claimed in the Icelandic debates that the health sector database would only contain information from medical records (no genetic or genealogical data) (see Árnason and Árnason 2004), which, according to deCODE literature (Íslensk erfðagreining 1998, p. 10), is not the case. Second, the different kinds of data have very different privacy requirements. Genealogical information is considered public information in Iceland; no privacy restrictions apply to it and it is frequently published in books and newspapers. Still, in order to keep individuals unidentifiable, genealogical information will only appear in query results as degree of kinship, not as family trees. Genetic data, however, is considered private in Iceland just as in other countries and, it was claimed, it would only be entered in the database if the individual has given informed consent. Medical data will be entered in the database on the principle of presumed consent.

There are at least two ethical worries here. One is the claim that no one can give informed consent for providing genetic data for the database as there are no specific research plans available. It is not possible to know how exactly the genetic data will be used or what risks it may involve. The second worry is a law on tissue sample banks (keeping samples of, for instance, blood or tissue), which was quietly passed by the Parliament in March 2000. This law proposes somewhat paradoxically both (1) that tissue samples collected for specific scientific studies can only be kept after the study or passed on to a tissue sample bank if the provider of the sample gives informed consent; and (2) that samples collected for clinical purposes (for example, diagnostic tests), can be passed on to tissue sample banks as long as the provider of the sample does not object (presumed consent). According to one Member of Parliament, this arrangement was a compromise between conflicting interests, that of individual privacy and that of freedom of science:

> The situation may arise that the individual does not want his sample to be used but that it is very important in order to find something out related to a specific study which can bring progress to the human kind and be for the good so there is a bit of a conflict. The conclusion was to do it this way, to split it up and have only presumed consent for [passing to tissue sample banks samples obtained from] clinical tests. (Hlöðversdóttir 2000, my translation)

The idea seems to be that if someone refuses to contribute a tissue sample to a study, it will be possible to obtain her sample from a tissue bank—as long as she has given a tissue sample for clinical purposes and not refused to have it passed on to a tissue sample bank. In other words, if you are asked for a tissue sample you can say no, but the researchers will most likely get your sample anyway, unless you have taken extreme care to prevent any tissue samples taken during service research from being passed on to tissue sample banks.

As the data is to be "non-personally identifiable," informed consent is not required from the individual when his or her medical data is transferred to the database. It is, however, possible to opt out by signing a non-consent form, which will prevent any future information about the individual being entered in the database—once information is entered in the database it is not possible to erase it. Parents can sign non-consent forms for their children, but many of those most vulnerable of adults who are not able to access or understand the information or tend to paperwork, because of mental illness, drug addiction and so on, would, as a matter of course, have their medical information entered in the database. The Directorate of Health keeps a list, or a database, of those who have opted-out of the HSD. As one of deCODE's most vocal critics, historian of science Skúli Sigurðsson, has put it in personal communication, the Health Sector Database does not exist, but the Directorate of Health has already established a database of the "deviants" who will not allow their medical data to be used in the HSD.

Once the health sector database would be constructed, it is possible to query it for "statistical information on health, disease and treatment" (Íslensk erfðagreining 1998, p. 5) but it yields no information about single individuals or groups of fewer than 10 individuals. Like an oracle, it will merely answer the questions posed to it. And the questions must be asked in the right way: deCODE planned to run a consultancy service, helping clients to formulate questions and interpret the answers.

In this chapter I will be almost exclusively concerned with the genetic uses of the databases. It is often pointed out, correctly, that the health sector database was only partially meant for genetic research, but Kári Stefánsson and deCODE company literature make it very clear that genetic research is the main purpose of the database. Its purpose seems occasionally to go even beyond the scientific purpose of conducting genetic research, to a thorough geneticization of medicine (Lippman 1991). In a paper presented at the first IFCC-Roche Conference in Singapore in March 1998,[3] Kári Stefánsson said that the "ultimate goal of the database [is] to usher in an era of preventive health care and individual-based disease management practices based on human genetics" and a little later

> The goal of the [health sector] database is to bring the genetics together with medical phe-
> notypes and outcomes to create a totally informative population with which to search for
> drug targets and to model both disease and host-drug interactions. (Gulcher and Stefánsson,
> p. 526)

Human genetics, as a "discourse of truth," was deployed for the purpose of gathering support and preparing the way for the establishment of the Health Sector Database, in a conjunction with a discourse of "progress" and "benefits for human kind." Its authority, as a scientific discourse, is used to manage the Icelandic population, as a research population—as a *totally informative* research population. The genetics

---

[3]IFCC is the International Federation of Clinical Chemistry and Laboratory Medicine; Roche is Roche Diagnostics, a Hoffman-La Roche subsidiary.

discourse is one of the main sources for the success HSD proponents have had in subjugating most of the Icelandic population as docile and happy subjects of genetic science. The following two sections focus on eugenics as one of the historical roots of the genetics discourse, and on how the modern genetics discourse turns the Icelandic population into a scientific resource and a commodity, as a totally informative population.

5.3   Eugenics, Nationalism, Nazi Experiments and the Icelandic Health Sector Database

> Anyone who can mention Nazi experiments and this database in the same sentence does not deserve to live in Iceland. (Kári Stefánsson, CEO of deCODE genetics Inc., quoted in Specter 1999, p. 51)

The history of human genetics cannot be separated from the history of eugenics, as has often been pointed out,[4] but it may come as a surprise that the Icelandic Health Sector Database has an ancestor in the age of eugenics. In this section I will first look at the history of eugenics, both generally and in the particular case of Iceland, and its relation to nationalism and racism. I will then end this section with a brief discussion of the eugenic ancestry of the Health Sector Database.[5]

My discussion of eugenics, and much of the discussion in this chapter in general, is framed by Foucault's discussion of biopolitics even if it is not always made explicit. Foucault discussed racism (although he barely referred to eugenics at all) in his lectures at the Collège de France in 1976 (Foucault 2003b [1976]) and very briefly in his lectures in 1975 (Foucault 2003a [1975]), but not in a way that is any immediate help for analysing the rise of eugenics and its relation to racial hygiene, science, and the biopolitical management of populations (see Macey 2009).

The history of modern eugenics begins in 19th century England with Francis Galton, a cousin of Charles Darwin. It was Galton who gave eugenics its name and founded it both as a science and as a social agenda. Ideas about breeding a better race or stock of humans are certainly older than that. Plato, for instance, wanted marriage, sex and procreation to be so arranged in his ideal society that the best men and women would have many children while the inferior men and women had few children. Children of the best would be taken care of by state nurses and teachers, but children of the inferior parents as well as deformed children would not be raised

---

[4] Daniel J. Kevles has done more than most to remind us of the eugenic inheritance of human genetics, for instance in his book *In the Name of Eugenics: Genetics and the Uses of Human Heredity* (Kevles 1985); but see also Kevles (1992) and Kevles (1999).

[5] The following discussion is largely based on four studies: Kevles (1992), Kevles (1999), Karlsdóttir (1998a b). There is no lack of literature on the history of eugenics, good examples include Bashford and Levine (2010), Adam (1990), Paul (1995), and, with the focus on the German context, Weikart (2004).

at all, that is, they would be exposed or simply killed.[6] Plato justifies this policy by referring to the breeding of animals, suggesting that humans can be bred to perfection no less than horses. It is safe to assume that the idea of breeding better humans is as old as the practice of breeding animal stock, but this idea did not become embedded in scientific discourse and social policy until late 19th century.

Francis Galton's main eugenic writings are *Hereditary Genius* (1869) and *Natural Inheritance* (1889). Galton, like Plato, pointed to the breeding of animals arguing that human stock *could* be improved, and he pointed to the danger of degeneration arguing that human stock *should* be improved. Galton not only coined the term eugenics, but also *positive* and *negative* eugenics, and *desirables* and *undesirables*. Positive eugenics is the breeding of *desirables*, for instance encouraging people of good stock (healthy middle and upper class people) to have more children. Negative eugenics is the prevention of *undesirables* (the poor and the sick, the "feebleminded," alcoholics, prostitutes, criminals and so on) from procreating, in order to breed out bad characteristics. The problem was that the undesirables were perceived to reproduce at a higher rate than the desirables, which was believed to lead directly to social degeneration.

The main reasons for the rise of eugenics are commonly taken to be Darwinism, especially Herbert Spencer's social Darwinism, and social progressivism. Other important factors are the rise of statistics and its application to social policy and planning, as well as a concern with hygiene: the lack of hygiene as a social problem and the encouragement of hygiene as a social policy. The concern with health is now mostly focused on exhibiting recent advances in the medical sciences, not least in genetics, while the concern with hygiene is literally history. Eugenics is firmly at the center of new technologies of biopower, which emerged in the 19th century, in particular that pole of biopower, which Foucault called "biopolitics of the population" (Foucault 1978, p. 139; Foucault 2003b [1976], pp. 242–254).

Galton did considerable work in statistics and his follower and colleague Karl Pearson is one of the most important figures in the history of statistics.[7] The hygienic aspect of eugenics seems to be underestimated in most accounts. Eugenics as a social policy was closely tied to the rhetoric of hygiene and many eugenic institutions and journals had "racial hygiene" or "biological hygiene" in their names rather than "eugenics."

During the first two decades of the twentieth century the new science of eugenics prospered and spread around the world. Francis Galton established The Eugenics Education Society (later renamed The Eugenics Society) and the journal *Eugenics Review* in 1907. When he died in 1911, Karl Pearson took over the leadership of the eugenic movement in Britain. Scientific institutions for eugenics were also established in Britain at this time, most notably the Galton Laboratory for National Eugenics at University College in London, led by Karl Pearson. The Galton Laboratory published on of the most important journals in the field: *Annals of Eugenics*.

---

[6]Plato lays out his plans in book V of the *Republic*, see in particular 459A–461B.
[7]For more about Karl Pearson and the history of statistics see Hacking (1990).

In the United States the Eugenics Record Office was established at the beginning of the twentieth century under the leadership of Charles B. Davenport and it soon became a part of the biological research facilities in Cold Spring Harbor in Long Island, New York, with generous support from the Carnegie Institution of Washington (Kevles 1992, p. 6). In 1903 the American Breeders' Association was established in the United States. Soon a eugenics department was founded, and as the interest in eugenics and genetics increased the name of the association was changed to the American Genetic Association and its journal's name from *American Breeders' Magazine* to *The Journal of Heredity* (Popenoe and Johnson 1933, p. 348). The first journal devoted to eugenics was the German *Archiv für Rassen- und Gesellschafts-Biologie*, which was founded in 1904. The following year the Society for Racial Hygiene was established in Berlin. The scientific institutions came a little later in Germany than in England and the United States: in 1923 a chair for race hygiene was founded at Munich with the appointment of the biologist Fritz Lenz. And in 1927 the Kaiser Wilhelm Institute for Anthropology, Human Heredity, and Eugenics was established in Berlin. Its first director was the anthropologist Eugen Fischer who was then the head of the Society for Racial Hygiene.

In 1912 a French eugenics society was established, its journal was named *Eugénique*. Brazilians followed in 1917, establishing in Sao Paulo a eugenics society which published *Annaes de Eugenía*. In Moscow a eugenics society was established in 1920, publishing *Russkii Evgenicheskii Zhurnal*. The first international eugenics conference was held in 1912 in London, drawing 700 participants. By 1924 the international eugenics organization had 15 member countries from Europe and the Americas, including the U.S., Britain, Russia, France, Germany, Italy, Norway, Denmark and Sweden. Other countries also had relations to the organization, including Canada, China and South-Africa.

British and, until the early thirties, German eugenicists were mostly worried about the lower classes, their proliferation in the cities as peasants migrated there in search of work, and a perceived increase in poverty, crime, alcoholism and prostitution. American and Canadian eugenicists were more worried about races and increased immigration from southern and eastern Europe (Kevles 1999, p. 436). It was only with the rise of National Socialism in the early thirties that German eugenicists became worried about the white race and its perceived decline. Such worries had been in the background in the eugenic rhetoric in northern Europe, but in National Socialism they came to the foreground.

Positive eugenics never really materialized in social policy, except for some relatively minor attempts in the United States at encouraging the good to breed more, for example through "Fitter Family" competitions often held at fairs in the 1920s. These competitions took place in the "human stock" section of the fairs. At the 1924 Kansas Free Fair families could compete in three categories, small, average and large; the winning families received a "Fitter Family Trophy" from the state governor and "Grade A Individuals" received a medal (Kevles 1992, p. 10; Kevles 1999, p. 436).

Negative eugenics were, on the other hand, implemented as a social policy primarily in the United States and later in Europe, by way of eugenic sterilization laws. The 1927 U.S. Supreme Court decision of *Buck versus Bell* declared such sterilization laws constitutional, and by then about two dozen states had passed eugenic sterilization laws. California was a leader in sterilization: by 1933 more people had been sterilized there than in all the other states combined (Kevles 1992, p. 10). Special targets of sterilization were the "feebleminded" and "morons," that is, the intellectually disabled, most of whom were in care of state institutions.

The German eugenicists were impressed by U.S. eugenic policies, particularly as they were implemented in California. During the early thirties, however, eugenic policies implemented by the National Socialists surpassed the American ones. One American eugenicist complained that "The Germans are beating us at our own game" (Karlsdóttir 1998b, p. 116). Eugenics was taken to its horrendous extreme in Nazi Germany, as is well known, with massive sterilizations and eventually death camps and racial extermination. Nonetheless, as the present discussion should make clear, eugenics cannot be equated with Nazi medicine. Eugenics movements existed in most Western countries and eugenic policies were implemented in many of them.

In the thirties, as eugenics became a more explicit part of the German Nazi rhetoric, it began to meet resistance in many countries, including the United States, Britain and the Soviet Union. In the Soviet Union eugenics was rejected primarily because of its association with Nazism. In the United States and Britain scientists turned against eugenics both because of its association with the Nazis and, according to Kevles (1992, p. 11), because it was considered scientifically poorly justified and socially and racially biased. Another probable reason is the lack of fruitful research in eugenics. After initial enthusiasm, the enormous collection of material and considerable speculation, not much in terms of actual scientific research took place. The classifications frequently used by eugenicists did not help. Davenport, for instance, studied the inheritance of such traits as "nomadism," "shiftlessness," and "thalassophilia"; the last being "the love of the sea," to quote Kevles, "that he [Davenport] discerned in naval officers and concluded must be a sex-linked recessive trait because, like color blindness, it was almost always expressed in males" (Kevles 1992, p. 11). Nowadays we have studies of the genetics of "adventurousness" and "novelty-seeking."

Eugenics did not fall entirely out of favour outside of Germany during the thirties. As a reaction to the criticism of the old eugenics, some scientists who studied human heredity and had eugenic inclinations, proposed what was termed "reform eugenics," which was meant to have stronger scientific underpinnings, be more humane and avoid class and race bias. The eugenic sterilization laws in the United States, however, were passed in the twenties in the context of the old eugenics. One of the most influential books in this area was *Sterilization for Human Betterment* by American eugenicists Ezra Gosney and Paul Popenoe, which was published in 1929. It discussed the "success" of Californian sterilization laws, and was a considerable authority in discussions of sterilization laws in other countries (Gosney and Popenoe 1929).

The Nordic countries soon followed the United States: sterilization laws were passed in Denmark in 1929, Norway and Sweden in 1934, Finland in 1935 and Iceland in 1938 (Karlsdóttir 1998b, p. 117). The discussion of these laws in the Nordic countries was more in the context of reform eugenics, at least among scientists and policy makers, although the public discussion tended to be more in line with the old eugenics. Eugenic arguments, for example worries about degeneration, were part of the justification for sterilization laws, but they were also justified through economic arguments and humane/paternalistic arguments (namely that sterilization is in the "poor idiots'" best interest). The sterilization laws were not widely debated, however, and they were passed with little or no objections; the eugenic arguments went generally uncontested.

When the nightmarish horrors of Nazi eugenic policies and practices in the extermination camps became widely known in 1945, the eugenics movement dissolved but eugenic ideas did not. Many eugenicists kept arguing for increased reproductive control in order to fight hereditary "flaws," but they distanced themselves from the racism, classicism and human breeding schemes of the old eugenics. Sterilization laws with explicit reference to eugenic purposes remained at least in the Scandinavian countries until the seventies.

Iceland did not have a eugenics movement, partly because there were little class or racial tensions and limited urbanization, and partly because the number of academics was too small for an academic movement of any sort. Still, eugenic ideas did have a number of proponents in Iceland. In January 1913, the district director of health in Akureyri, dr. Steingrímur Matthíasson, held a public lecture about eugenics with the title "The Decline of Our World." He claimed that foreign scientists were growing worried about the decline and degeneration of the white race, caused by unhealthy living, corrupting cities and the increased number of people who were saved from natural selection by modern medicine. The doctor believed the Icelandic population was starting to show the signs of degeneration, and he feared that this development could not be reversed merely by medicine and hygiene: "Society must make laws which prevent the sick and the miserable from filling the Earth. Eugenics and hygiene must join forces and show us this right direction" (Karlsdóttir 1998a, p. 425, my translation).

The most vocal proponents of eugenics in Iceland were not medical doctors, but two philosophers at the University of Iceland, Ágúst H. Bjarnason and Guðmundur Finnbogason. Ágúst H. Bjarnason, like the medical doctor in Akureyri, was particularly worried about how medicine and health care institutions were interfering with the law of natural selection. In the second volume of his book on ethics, published in 1926, he writes:

> Now it is, for instance, considered a clear duty of mercy to care for people who are sick and weak and may perhaps not survive at all. That is fine; it is just nice. But when those same people are allowed, without hindrance or objection, to procreate and hence corrupt the stock, then it becomes more than questionable, whether this is morally correct and permissible, and one starts asking oneself, whether the law should not prevent this. And if man does not do this, then we can expect only that nature will, that sooner or later nature wipes all that is sick and corrupt off the face of the earth. (Karlsdóttir 1998a, p. 427, my translation)

In this example of eugenic writing, one finds the two poles so common in eugenicist writings in general, on one hand a strong faith in progress, on the other hand an almost apocalyptic pessimism.

In 1921 Guðmundur Finnbogason published a book on the Icelandic nation which was heavily influenced by the eugenics movement, and the following year he published an article in which he complained that eugenics had not been given much consideration in Iceland. His article was intended to educate the Icelandic public about this new and important science and show what Icelanders could learn from it (Karlsdóttir 1998a, p. 427). In his article he discussed favourably the claims made by British eugenicists that poverty, criminality and misery are inherited characteristics, and then happily pointed out that in Iceland there were no inherited class differences. Quite the contrary, he considered the Icelandic nation as a whole to be of excellent stock. He claimed that the original stock (of Norse and Celtic settlers in the ninth and tenth century) was particularly good, because only the strongest and most diligent men would have travelled this far only to face the dangers and hardships of founding a settlement on a distant island. Since then the stock had only improved, first because almost no foreign blood had mixed with the blood of the original settlers and second because the hardships over the centuries (eruptions, brutally long and cold winters, plagues and so on) had weaned out the weakest and so made the stock even stronger (Quoted in Karlsdóttir 1998a, pp. 79–80). Both themes reoccurred in the Database debate, as I will discuss below.

Unlike Britain, class differences were not important for eugenicists in Iceland, but like in the United States and Nazi Germany, eugenic ideas in Iceland were from early on mixed with nationalism and racism. Eugenics, nationalism and racism stem from the same roots in 19th century epistemic, cultural and social developments, and could easily mix with and reinforce each other.

Racism has been a difficult subject in Iceland. During the Second World War Iceland refused to take Jewish refugees and expelled some who had already fled to Iceland. The reasons are not just plain racism, they also get their force from the authority of science, as is apparent in one newspaper editorial from 1938 with the title "Protecting the Race":

> No one is in as much danger as small nations like Iceland from the immigration of foreign men and the resulting mixture of the race. An important Icelandic scientist has said that only fifty Jews were required for the nation to lose its Nordic characteristics in 2–3 generations. (quoted in Karlsdóttir 1998b, p. 103, my translation)

The author of the editorial wrote that he understood the Jews' need for refuge, but it was the responsibility of the big nations to solve that problem. The author was worried because Jewish refugees were already entering the country and he demanded action from the authorities before it would be too late (that demand was not necessary, the authorities were already working on this "problem"). The editorial went on:

> And the nation has [...] the holy duty to protect the Icelandic race, the Nordic and Celtic blood, in order to avoid its mixture with a strong foreign race, which can wipe out the Nordic family characteristics in a few generations. It must be an unconditional demand of every Icelander, that the government take care to strongly hold back the immigration of foreigners, who now seek refuge all over Europe. (quoted in Karlsdóttir 1998b, p. 104, my translation)

The government was of the same conviction as the author, although the rationale for denying Jewish refugees entry to the country was less racist. The official justification was that the high unemployment rate and common poverty in Iceland did not leave any room for refugees, but it is was quite clear that this was a pretext. Even wealthy

Jews who requested merely permission to stay in Iceland, without a work permit, were denied entry (Karlsdóttir 1998b, p. 104).

The history of eugenics in Iceland has an interesting parallel to the techno-scientific database discourse in Iceland. When proponents of the database discuss the value of the Icelandic nation for genetic research they typically point to three things: Extensive genealogical information, good medical records and the genetic homogeneity of the nation. The most important of the three is the nation's perceived homogeneity: Icelanders are taken to be more (genetically) alike than other nations. Not only was the population quite isolated from the eleventh century until the Second World War, but it went through three major "bottlenecks," periods when disasters killed off a large part of the population. In the early 15th century the plague killed around one third of the approximately 70,000 inhabitants of Iceland, in 1707–1708 over one third of the population died of smallpox, and in 1783 Lakagígar, a row of volcanoes in the south of Iceland, erupted, resulting in the death of a quarter of the population. By 1800 Iceland had only 45,000 inhabitants, on January 1, 2018, they were 348,450 (according to Statistics Iceland 2018).

These population bottlenecks are claimed to have decreased the variation of geno-types in the population. As the metaphor goes, the Icelandic gene pool is shallow. This is expected to facilitate genetic research, because when genotypes of rela-tively homogenous individuals are compared there are fewer variations in genotypes (genetic characteristics) that could account for observed variations in phenotypes (physical characteristics).

The observation that the Icelandic nation has been subjected to more drastic natural selection than most other nations, and that this somehow makes the nation valuable for science, is not as new as many believe. In the twenties and thirties, Ice-landic eugenicists noted what is now called bottlenecks in the history of the nation. This made the nation valuable, not of course because of homogeneity but because the stock had been improved through "a thousand hardships" (Karlsdóttir 1998a). Around this time ideas appeared about Iceland's role in the future of eugenics. It was pointed out that Iceland had thorough genealogical records, which would be par-ticularly useful for eugenic research. One of the philosopher-eugenicists mentioned above, Guðmundur Finnbogason, proposed in 1922 that a genealogical institute be established, which would accumulate genealogical information and put together a card catalogue, a database of sorts, where all Icelanders would be registered along with information about their health and hereditary characteristics, their biographies and genealogies, as well as samples of their hand writing and voice, and anything else that could serve as the basis for research in genealogy and heredity.

> Scientists would work there, who would want to investigate the heredity of particular char-acteristics, mental and physical, and those men would go there who would be interested in getting to know the nature of some specific families. There one could see if the race was changing for the better or worse. There would be at every time a looking glass, showing how the heart of the nation beats. (quoted in Karlsdóttir 1998a, p. 58, my translation)

Guðmundur Finnbogason described the importance of such an institution in words that could be taken out of the database discourse: "A great task awaits Icelanders.

They should become and could become that nation which lays down the widest and most solid foundation for the hereditary research of the future" (quoted in Karlsdóttir 1998a, p. 59, my translation).

The other philosopher and supporter of eugenics mentioned above, Ágúst H. Bjarnason, discussed these ideas in 1926 and thought they would be rather expensive to realize and therefore not feasible at the time. Until then, he suggested that all Icelanders individually record their genealogy and information about themselves and their relatives "and hide neither virtues nor faults, although they concern close relatives" (quoted in Karlsdóttir 1998b, p. 59, my translation).

During the war, in 1943, support for the idea came from Halldór Laxness, Iceland's greatest writer and social critic and winner of the Nobel prize for literature in 1955. In his article "Human Life on a Card File," Laxness suggests that information about Icelanders could be registered on cards and organized, and then sold to clients for an appropriate fee, although his motivation came not from eugenics but rather from his unhappiness with the fact that only the lives of society's "important men" were being registered. Still, the information in the Icelandic "card file" was to be much more than just family relations and biographies. Laxness was aware of efforts by the German Nazis to construct databases with such personal information, and distanced himself from their aims, but not from the basic idea of collecting, organizing and making use of such data:

> I am told that the German secret police has a card file containing information about millions of men from all over the world, with comments on their origin, behavior, views and character, as well as biographical information—all for the purpose of being able to walk up to these men and kill them at the opportune moment. Compared to Himmler's card file, which is made for murdering, it would be easy to give these few Icelanders life on a card file. (Laxness 1946 [1943], pp. 178–181, my translation)

Actual plans for a genealogical institution and a database for the Icelandic nation for hereditary research were not made at this time, due to lack of funds and lack of political and scientific interest. In the discussion about this possibility, many of the themes in the present database discourse occur: the basic idea of establishing a database with hereditary and health information about all Icelanders, the possibility of selling this information, the idea of observing an entire population—"how the heart of the nation beats" in the words of Guðmundur Finnbogason—the value of the extensive genealogical records, as well as the eugenic idea of a strong, pure and unspoiled race, a homogenous population in the terms of modern genetics, which has only become purer and stronger through the hardships of the centuries, the "population bottlenecks" in the language of modern human genetics.

5.4 The "Totally Informative Population"

The eugenicists believed that the Icelandic "stock" was valuable for scientific research, because of its noble Nordic and Celtic origin and merciless but ultimately beneficent circumstances through its history. Modern geneticists consider the nation valuable because of its homogeneity, which also points to the nation's Nordic-Celtic origin (small founding population) and its often horrific history in Iceland (in particular the population bottlenecks). If we look to the past we can see historical and rhetorical parallels of the homogeneity of modern genetics and the good stock of eugenics, but, looking to the present, homogeneity points to laboratory animals. Laboratory animals, in particular laboratory mice and rats, are bred specifically for the laboratories and do not exist in the wild. Experiments require the elimination or at least reduction of such distorting factors as the peculiarities of the individual organisms. A part of a controlled experiment is to produce an ideal situation, where the phenomenon being investigated can appear in its purest form. Much progress in transmission genetics (that is, the study of heredity) has emerged from the study of one specific life form: *drosophila*, the fruit fly. One of the central problems in the early *drosophila* studies around 1910 was to breed and maintain the right stock of flies. Since not just any fruit fly would do for these studies, geneticists shared fruit flies of "the right stock," otherwise the experiments just would not work. As a result, a specific type of *drosophila* came in existence, bred particularly for genetic studies (Kohler 1999).

It may seem an exaggeration to compare a nation with such stocks of laboratory animals, but a look at articles on the database that have appeared in the popular media shows that this comparison is frequently made. The critics, of course, like to picture Icelanders as the new guinea pigs of genetics. On October 3, 1998, the Swiss news paper *Der Tagesanzeiger* published in its weekend supplement *Das Magazin* an article with the title "The Icelanders, Our Lab Mice" (Die Isländer, unsere Labormäuse), criticizing the database project (Fischer 1998). A front page article in the *Washington Post* on 12 January 1999 quotes David Banisar of the Electronic Privacy Information Center in Washington saying: "'Turning the population into electronic guinea pigs' should serve as a warning" (Schwartz 1999b).

Not only critics made the comparison with laboratory animals; scientific and financial magazines made it too and with curious lack of concern. The respectable *Scientific American* published an article in the February issue 1998 entitled "Natural-Born Guinea Pigs: A start-up discovers genes for tremor and psoriasis in the DNA of inbred Icelanders," which opens with these startling words:

> To build a life among the glaciers and volcanoes of Iceland takes a special breed of people. Not just figuratively, either: the 270,000 citizens of this island nation, a great majority of them descended from seventh-century Viking settlers, form one of the most inbred populations in the world. (Gibbs 1998, p. 34)

One investor, Brian Atwood of Brentwood Venture Capital, doubted the usefulness of the Icelandic nation for genetics in an interview with the business magazine *Red Herring*, going further than just making a comparison of Icelanders to laboratory animals: "Despite the high degree of inbreeding within the Icelandic population, Atwood thinks that the amount of *inter*breeding is still high compared with that of mice or fruit flies, which can be bred more narrowly" (*Red Herring* 1998). It is irrelevant that Atwood was speaking as an investor in a rival genomics company and hence quoted as a financial critic of deCODE, the point is that a financial magazine not only compares the nation to lab animals but—as if it was the most normal thing in the world—speaks of a nation *as* lab animals, as laboratory humans that can be bred for experiments, albeit not as narrowly as fruit flies.

deCODE itself has also presented the Icelandic population in a way which suggests laboratory animals. In an information brochure for potential investors deCODE boasts that "deCODE Genetics will be in a position to offer its corporate partners access to the Icelandic population for clinical trials of drug candidates" (Árnason and Árnason 2004, p. 171n). Icelanders are represented as laboratory animals, which makes Iceland the laboratory: "The genetic similarity might make Iceland the best laboratory in the world to study genes" says an ABC News story (Gizbert 1999); and in a story on genetic research in Newfoundland, Toronto's *Globe and Mail* writes: "a province-sized laboratory has a conceptual precedent: Iceland" (Atkinson 2000). The walls of the laboratory have been torn down and now it can encompass a Canadian province or an island in the Atlantic. In the new world of genetics, the laboratory animals are no longer locked in cages in the laboratory: they are us.

As laboratory animals, the Icelandic nation is very valuable according to both geneticists and investors. The geneticist Mary-Claire King, professor in the Department of Medicine and Genetics at the University of Washington, marvelled: "Iceland is just an amazing place to do genetics. The population there is like a gift from heaven" (Specter 1999, p. 42). And Elizabeth Silverman of BancAmerica commented in *Nature Biotechnology* on a US$ 200 million deal between deCODE and Swiss pharmaceutical company Hoffman-La Roche that "the nation is extremely valuable" (Dorey 1998).

The scientific and commercial value of Icelanders consists in the perceived homogeneity of the nation. The international media discussed Icelandic homogeneity in terms that should surprise anyone who has walked around in Iceland with open eyes. Specter's *New Yorker* article is fairly typical:

> It's a cliché, but the first thing a visitor to Iceland notices, after the volcanic landscape that lies beneath the approach to Keflavik International Airport, is just how closely Icelanders resemble each other [...] Iceland sometimes seems to be inhabited by one enormous family, not one of whose members ever leaves the neighborhood where he was born. [... T]he hereditary instructions for blue eyes and blond hair, which are so prevalent in Iceland, have been passed undiluted through a small gene pool for fifty generations. After a thousand years of plagues, epidemics, earthquakes, and volcanoes finished weeding out the population, what remains [...] is a nation of two hundred and seventy thousand of the most genetically similar people on earth. (Specter 1999, pp. 41–42)

The idea, myth even, of Icelanders' homogeneity was constantly reworked in the media. Let me give just three more examples. First, John Schwartz, journalist at the *Washington Post*, writes: "The strikingly uniform DNA of Iceland's largely blue-eyed, blond-haired populace is expected to provide an invaluable resource for studying human genetics, leading to fundamental insights into many diseases, proponents say." A few paragraphs below this myth of a homogeneously blond and blue-eyed population is repeated: "Iceland's population presents a tantalizing opportunity for those who study genetics because Icelanders' blond hair and blue eyes reflect one of the most remarkably homogeneous populations in the world." And he adds: "The original blend of 9th century Norse stock and Celtic seamen has been largely unchanged, and that gene pool was further restricted by bouts of plague, famine, and volcanic eruption" (Schwartz 1999b). Second example: Novelist Simon Mawer wrote in an op-ed article in *The New York Times* entitled "Iceland, the Nation of Clones": "But since everyone in Iceland is related to everyone else there (all of them are descended from the same few Vikings), the place is, comparatively speaking, a nation of clones" (Mawer 1999). Third example: A news story on the ABC News web site shows a picture of a blond girl (presumably Icelandic), with a caption that reads: "Virtually all of Iceland's 270,000 residents trace their roots to the Vikings. Their homogeneity might make Iceland the best laboratory in the world to study genes" (Gizbert 1999).

deCODE's own research suggests slightly less genetic difference in the Icelandic population than in ten European populations: "the heterozygosity rate [of microsatellite markers] over 300 Genethon markers in Iceland is 0.75 compared with 0.79 in Europe (Icelandic data is unpublished)" (Gulcher and Stefánsson 1998, p. 524; see also Gulcher et al. 2000, p. 395). But their main arguments for homogeneity are not genetic but historical and linguistic, referring to the small founding population (although no one knows how small it was, likely at least a few thousand and possibly tens of thousands) and subsequent isolation and population "bottlenecks" over the following one thousand years, as well as little change in the Icelandic language from the 13th/14th century (when the Sagas were written) to this day (Gulcher and Stefánsson 1998, p. 524). All of these arguments have been challenged, primarily on historical grounds. As for the scientific evidence, one study notes that "claims about special genetic homogeneity of Icelanders are not supported by evidence," and concludes: "Examination of the published literature on blood group and allozyme variation does not provide any support for the notion of special genetic homogeneity of the Icelanders. Further studies of microsatellite variation are unlikely to do so" (Árnason 2003, pp. 5, 14; see also Árnason et al. 2000, and Árnason 1999).

It is not important that a minority of Icelanders are blond, nor that brown eyes are quite common, nor that Icelanders are not inbred, nor even that no independent scientific evidence exists for the claim that the Icelandic population is homogeneous. What is interesting is that this claim, that Icelanders are homogeneous, is useful for both business and science, and that this claim—and its justification—is practically the same as the one made by eugenicists in Iceland in the 1920s and 1930s: Icelanders are a very special stock because of their Viking origin and "good breeding through a thousand hardships."

The argument here is not that proponents of the Health Sector Database are wrong or immoral because they are arguing for views, which Nazis or eugenicists advocated or would have applauded; that would merely be guilt by association, a *reductio ad Hitlerum* (Strauss 1953, pp. 42–43). The argument is that proponents of the Health Sector Database sought to give their claims force through a combined source of science, medicine, and a nationalistic, even racist, mythology and thought-style, and that this source of power was also used by eugenicists almost a century ago. This source is dubious, not because it was used by eugenicists, but because it contains and reinforces myths, stereotypes and thought which is at bottom nationalistic and racist.

When one understands what the source of power is, and how quickly it evaporates when put to critical analysis, the claims lose their power-effects.

Icelanders are not the first group to be singled out for genetic research, similar fate has befallen Ashkenazi Jews, American Mennonite communities, Sami people in Finland and the population of the Atlantic island Tristan da Cunha, among others. But this is the first time that an attempt is made to turn an entire nation into a constant source for genetic research, a perpetual laboratory population, a nation in a petri-dish. The way the population is represented, as laboratory animals, reflects the subjugation of the population as docile research subjects, which in turn is one of the power effects of the techno-scientific database discourse. It is a power effect which the scientific subjects, the Icelanders, have been more or less unable to resist. Identifying and analyzing the database discourse, and pointing out its effects, is, I believe, a step towards resistance. It is a critique, it is a politics of truth.

Politics of truth, as a type of critique, is at least partly directed at what Foucault called "the modern régime of truth," as I discussed in Sect. 3.4. The main features of the modern régime of truth could be identified where left-handers become subjects of science (see Sect. 4.4), but I have not pointed out these relationships in the case of Icelanders as subjects of science. Let me therefore give a brief overview.

Knowledge about Icelanders *is centred on the form of scientific discourse and the institutions which produce it*. Icelanders, as a population that becomes the subject of science, are described as valuable for genetics research and are defined in terms of their value for science. The population is not only compared to laboratory animals, like mice and fruit-flies, but is actually discussed as a population of laboratory animals and compared to other types of laboratory animals. Various strategies and techniques are deployed to manage the population as a scientific subject and to ensure its docility and cooperation. An interesting feature of this case is the deployment of the discredited scientific discourse of eugenics, along with the nationalist and racist discourse that permeates it, as evidence for the homogeneity of the Icelandic population. There is a curious tension between this discredited sort of evidence and the evidence from molecular genetics which appears to contradict the claim that the population is genetically homogenous.

Knowledge about Icelanders *is subject to constant economic and political incitement*. The Health Sector Database project was initiated by a private company and required the support of the government and the parliament. The main aim of the private company was to turn profit using the medical records, genetic data, and genealogical information of the population of Iceland as its resource in an environment of immense enthusiasm and optimism about human genetics. The economic incitement is therefore clear. The political incitement for this knowledge production is perhaps secondary, but an important part of the appeal of the HSD project was its expected use for public health, for a biopolitical management of the population that goes far beyond basic statistics about birth rates, mortality, and longevity. The project would not only create a "totally informative population" for genetics research, but also a population whose health status was transparent to both science and government, a public health panopticon.

Knowledge about Icelanders *is the object, under diverse forms, of immense diffusion and consumption.* The debates about the HSD project involved considerable discussion about the scientific knowledge we have about the Icelandic population, the available knowledge was diffused and consumed both in Iceland and abroad. The aim of the HSD project was to create a resource for the production of knowledge about the Icelandic population and the business model was based on an expectation that this knowledge would be widely diffused and consumed. The debate itself took place both nationally and internationally in scientific journals, in popular science magazines, and in the mass media. The somewhat inexact image of the Icelandic population as homogenous, blond and blue-eyed and inbred, was endlessly reproduced as evidence for the value of the population for genetics research. Much of the knowledge of Icelanders that was circulated in the media was not strictly scientific, but it was circulated in the context of a scientific endeavour and as a part of the scientific rationale for the project.

Knowledge about Icelanders *is produced and transmitted under the control, dominant if not exclusive, of a few great political and economic apparatuses.* These apparatuses include at least the university, media, writing, and corporations. One corporation, deCODE genetics, plays a leading role in this case, but much of the knowledge deployed in the debates were produced within universities and circulated through media and writing. This case introduces an interesting tension between private corporate research and public university research. I discussed briefly the problem of scientific freedom and the exclusive rights that deCODE would have had to the medical data in the Health Sector Database.

Knowledge about Icelanders *is the issue of a whole political debate and social confrontation.* This chapter is largely about the political debate and the social confrontation regarding the Health Sector Database project. The plans themselves and the debates about them already influenced the image and identity of the Icelandic population both nationally and internationally. If the plans had been successful, the population would have been made both "totally informative" for genetics research and totally transparent for public health as a biopolitical management tool. There are various reasons for Icelanders to resist these power effects of this particular scientific discourse about Icelanders, to resist being made into this sort of a scientific subject. My hope is that the analysis of the discourses around this debate contributes to this resistance by criticizing both the scientific (and semi-scientific) discourses at work and their particular power effects.

In this chapter I have discussed certain central aspects of the scientific database discourse, in particular the human genetics discourse and its roots in eugenics, which is deployed to establish the authority of proponents of the database project. I have also discussed how the population of Iceland is subjugated and managed as a laboratory population. My discussion is in contrast to the conventional bioethical discourse, which all too often ends up being nothing more than a risk/benefit analysis. I have attempted to throw some light on what effect this techno-scientific and commercial project has on the Icelandic population just in virtue of being conducted, regardless of its possible application for good or bad ends. It is important to investigate ethical questions about how scientific knowledge is used, but there must also be investigations concerning the power effects of scientific knowledge and scientific research programmes. The human subjects of science can act as "local critics," as opposed to universal intellectuals. The subjects of science should not leave the politics of truth to the scientists.

References

Adam, Mark B. (ed.). 1990. *The Wellborn Science: Eugenics in Germany, France, Brazil, and Russia*. Oxford: Oxford University Press.
Árnason, Alfreð. 1999. Hvað með MS-genið og hina einsleitu þjóð? *Morgunblaðið* 12 March.
Árnason, Einar. 2003. Genetic Heterogeneity of Icelanders. *Annals of Human Genetics* 67: 5–16.
Árnason, Einar, Hlynur Sigurgíslason, and Eiríkur Benedikz. 2000. Genetic Homogeneity of Icelanders: Fact or Fiction? *Nature Genetics* 25: 373–374.
Árnason, Garðar. 2004. Interbreeding Within the Icelandic Population is High Compared to That of Mice or Fruit-Flies. In *Blood and Data: Ethical, Legal and Social Aspects of Human Genetic Databases*, ed. Garðar Árnason, Salvör Nordal, and Vilhjálmur Árnason. Reykjavík: University of Iceland Press/Háskólaútgáfan.
Árnason, Vilhjálmur, and Garðar Árnason. 2004. Informed Democratic Consent? The Case of the Icelandic Database. *Trames* 8: 164–177.
Atkinson, Bill. 2000. The Rush for The Rock. *The Globe and Mail* 5 January. http://www.theglobeandmail.com/technology/science/the-rush-for-the-rock/article18420308/. Accessed 14 Sept 2018.
Bashford, Alison, and Philippa Levine. 2010. *The Oxford Handbook of the History of Eugenics*. Oxford: Oxford University Press.
Chadwick, Ruth. 1999. The Icelandic Database—Do Modern Times Need Modern Sagas? *British Journal of Medicine* 319: 441–444.
Crosby, Jackie. 1999. Iceland: The Selling of a Nation's Genetic Code. *Minnesota Star Tribune* 10 Feb.
Dorey, Emma. 1998. Roche deCODEs Icelandic Population in $200 Million Deal. *Nature Biotechnology* 16 (13 March): 225–226.
Fischer, Heinz Joachim. 1998. Á kóðavegi og heim aftur. *Morgunblaðið* 28 October. http://mbl.is/greinasafn/grein/427826/. Accessed 14 Sept 2018.
Foucault, Michel. 1978. *The History of Sexuality I: An Introduction*. New York: Pantheon. Trans. R. Hurley. Originally *Histoire de la sexualité I: La volonté de savoir*. Paris: Gallimard, 1976.
Foucault, Michel. 2003a [1975]. *Abnormal: Lectures at the Collège de France 1974–1975*, eds. Arnold Davidson, Valerio Marchetti, Antonella Salomoni, Alessandro Fontana, and François Edwald. New York: Picador.
Foucault, Michel. 2003b [1976]. *"Society Must Be Defended": Lectures at the Collège de France 1975–1976*, eds. Arnold Davidson, Mauro Bertani, Alessandro Fontana, and François Edwald. New York: Picador.
Gibbs, William Wayt. 1998. Natural-Born Guinea Pigs: A Start-Up Discovers Genes for Tremor and Psoriasis in the DNA of Inbred Icelanders. *Scientific American* 278 (February): 34.
Gizbert, Richard. 1999. Profiling an Entire Nation. *ABC News*. A copy is available at https://web.archive.org/web/20031005215852/http://abcnews.go.com/onair/WorldNewsTonight/wnt990218_iceland.html. Accessed 14 Sept 2018.
Gosney, Ezra, and Paul Popenoe. 1929. *Sterilization for Human Betterment*. New York: Macmillan.
Greely, Henry T. 1992. Health Insurance, Employment Discrimination, and the Genetics Revolution. In *The Code of Codes: Scientific and Social Issues in the Human Genome Project*, ed. Daniel Kevles and Leroy Hood, 264–280. Cambridge, MA: Harvard University Press.

Greely, Henry T. 2001. The Revolution in Human Genetics: Implications for Human Societies. *South Carolina Law Review* 52: 377–381.

Gulcher, Jeffrey, and Kári Stefánsson. 1998. Population Genomics: Laying the Groundwork for Genetic Disease Modeling and Targeting. *Clinical Chemistry and Laboratory Medicine* 36: 523–527.

Gulcher, Jeffrey, Agnar Helgason, and Kári Stefánsson. 2000. Genetic Homogeneity of Icelanders. *Nature Genetics* 26: 395.

Hacking, Ian. 1990. *The Taming of Chance*. London: Cambridge University Press.

Häyry, Matti, Ruth Chadwick, Vilhjálmur Árnason, and Garðar Árnason (eds.). 2007. *Ethics and Governance of Human Genetic Databases: European Perspectives*. Cambridge: Cambridge University Press.

Hlöðversdóttir, Bryndís. 2000. Parliamentary speech, session 125, meeting 116, 11 May. http://www.althingi.is/altext/125/05/r11131356.sgml. Accessed 14 Sept 2018.

Íslensk erfðagreining. 1998. Gagnagrunnur á heilbrigðissviði. Spurningar og svör ["Health Sector Database. Questions and Answers"]. A free information booklet published by deCODE.

Kahn, Jeffrey P. 1999. Attention Shoppers: Special Today—Iceland's DNA. CNN Ethics Matters. http://edition.cnn.com/HEALTH/bioethics/9902/iceland.dna/template.html. Accessed 14 Sept 2018.

Karlsdóttir, Unnur Birna. 1998a. Kynbætt af þúsund þrautum. *Skírnir* 172: 420–450.

Karlsdóttir, Unnur Birna. 1998b. *Mannkynbætur*. Reykjavík: Sagnfræðistofnun Háskóla Íslands & Háskólaútgáfan.

Kevles, Daniel J. 1985. *In the Name of Eugenics: Genetics and the Uses of Human Heredity*. Cambridge, MA: Harvard University Press.

Kevles, Daniel J. 1992. Out of Eugenics: The Historical Politics of the Human Genome. In *The Code of Codes*, ed. Daniel J. Kevles and Leroy Hood, 3–36. Cambridge MA: Cambridge University Press.

Kevles. Daniel J. 1999. Eugenics and Human Rights *British Medical Journal* 319: 435–438.

Kohler, Robert E. 1999. Moral Economy, Material Culture, and Community in *Drosophila* Genetics. In *The Science Studies Reader*, ed. Mario Biagioli, 243–257. New York and London: Routledge.

Laxness, Halldór K. 1946 [1943]. Mannlíf á spjaldskrá. *Sjálfsagðir hlutir*. Reykjavík: Helgafell: 178–181.

Lippman, Abby. 1991. Prenatal Genetic Testing and Screening: Constructing Needs and Reinforcing Inequities. *American Journal of Law and Medicine* 17: 15–50.

Macey, David. 2009. Rethinking Biopolitics, Race and Power in the Wake of Foucault. *Theory, Culture & Society* 26 (6): 186–205.

Mawer, Simon. 1999. Iceland, the Nation of Clones. *The New York Times* 23 January.

McInnis, Melvin G. 1999. The Assent of a Nation: Genethics and Iceland. *Clinical Genetics* 55 (4): 234–239.

Paul, Diane B. 1995. *Controlling Human Heredity 1865 to the Present*. Atlantic Highlands, NJ: Humanities Press.

Popenoe, Paul, and Roswell Hill Johnson. 1933. *Applied Eugenics*. New York: The Macmillan.

Red Herring. 1998. Designer Genes: Icelandic Startup DeCode Takes a Homegrown Approach to Genomics. 11 May. http://money.cnn.com/1998/05/11/redherring/redherring_genes/. Accessed 14 Sept 2018.

Russo, Enzo, and David Cove. 1998. *Genetic Engineering: Dreams and Nightmares*. Oxford: Oxford University Press.

Schwartz, John. 1999a. Iceland to Make its Genetic Code a Commodity. *The Washington Post* 12 January.

Schwartz, John. 1999b. With Gene Plan, Iceland Dives Into a Controversy. *International Herald Tribune* 13 January.

Specter, Michael. 1999. Decoding Iceland. *The New Yorker* 18 January: 40–51.

Statistics Iceland. 2018. Statistics Iceland Website. http://www.statice.is. Accessed 14 Sept 2018.

Strauss, Leo. 1953. *Natural Right and History*. Chicago: University of Chicago Press.

The American Society of Hematology. 1999. The American Society of Hematology 41st Annual Meeting and Exposition. December 3–7. http://www.hematology.org/meeting/meeting99/special.html. Accessed 6 Dec 1999 (this file has been removed; a copy is available at https://web.archive.org/web/20000304140152/http://www.hematology.org/meeting/meeting99/special.html. Accessed 14 Sept 2018).

Weikart, Richard. 2004. *From Darwin to Hitler: Evolutionary Ethics, Eugenics, and Racism in Germany*. New York: Palgrave Macmillan.

Chapter 6
Conclusion

In this book I have attempted to develop a critical approach to science, which I call "politics of truth," and apply it to two cases, namely the scientific studies of left-handers and Icelanders. In way of conclusion, let me look back and draw together what I have done so far. Following the first chapter's introduction, I discuss in the second chapter Foucault's archaeology, somewhat as a prelude. I argued that Foucault only applied archaeology, as an explicit method, in *The Order of Things*, and, following a critical overview of Foucault's *Archaeology of Method*, I concluded that archaeology is poorly served by discussing it purely on a theoretical level. Archaeology stands and falls with its application in concrete historical cases.

In the third chapter I turned to developing "politics of truth" as a critical approach to science. I first discussed Foucault's concept of power, then looked at the relation between power and knowledge, arguing that although Foucault does not offer a general theory of power and knowledge, one can identify two basic kinds of relations between power and knowledge in his work. One has to do with the classification, labeling, and generally the studying of people, the other with biopower. An example of the first type of relation is how delinquents evolved, in the 19th century, as a category of people and as an object for study, along with the modern penitentiary apparatus. The second type of relation, biopower, has two main forms, which Foucault called *anatomo-politics of the human body* and *bio-politics of the population*. I concluded by pointing out that biopower is relevant to my discussion of life sciences and biotechnologies in so far as the disciplining of the body or the management of populations is involved.

In the third section of the third chapter, I argued that Foucault's discussion of power and knowledge could and should be extended from the sciences of man to the natural sciences. I showed that natural sciences have various sorts of power effects, including silencing and disqualifying people, both within and without science. The natural sciences can also determine, or at least affect, not only what sort of things one can say and think, but also what one can do and what one can be. I expanded on two examples, from Joseph Rouse and Hannah Landecker respectively, to show how this can happen. Rouse's example concerns how natural sciences provides a certain view

© The Author(s) 2018
G. Árnason, *Foucault and the Human Subject of Science*, SpringerBriefs in Ethics,
https://doi.org/10.1007/978-3-030-02813-8_6

of the world and our place in it, as well as an exemplary way of providing knowledge, not only of nature, but also of humans. Landecker's example concerns how people are turned into resources for biological materials and the struggle that can result from that between the scientists and the human biological resources themselves.

In the final section of the Chap. 3, I propose "politics of truth" both as a way of analysing specific power-knowledge relations in the context of science, and as a way of resisting the power effects of particular scientific programs or projects. I proposed politics of truth as an alternative to two conventional types of criticism of science: Epistemological criticism focusing on the concepts, methods and justification of science; and ethical criticism focusing on such issues as the ethical application of science or the treatment of human and animal research subjects.

To set a frame for "politics of truth," I discuss Foucault's idea of a "régime of truth" and his analysis of the modern régime of truth—where science is pivotal. In discussing Foucault's view on "truth," I offer an interpretation which takes "truth" seriously as something concrete and earthly, something that can be produced, diffused, and consumed; as something that has regular effects of power. According to Foucault, scientific knowledge functions as "truth" in the modern régime of truth.

Politics of truth, as an analysis, asks what economic and political forces create need and demand for the particular knowledge under consideration, how the knowledge is produced, transmitted, diffused and consumed; and how it gives rise to a political debate or social confrontation. Politics of truth, as a critique, has two objectives: one modest, the other immodest. The modest objective is to find ways of resisting the power effects of a specific scientific research program, partly by becoming aware of the subjugating elements of scientific knowledge and what one can do to influence the dynamics of power and knowledge in that particular case. The immodest objective is not only to defend oneself but to change the very relations between us and the régime of truth—or even the régime of truth itself. Foucault suggested the latter, changing the régime of truth itself, but I argued that this objective is rather too ambitious. I think that, more realistically, one may try to change the way people think about a given scientific discourse, in order to reduce or reshape the power effects it has; or one may try to change the way of thinking within science—to change how scientists think about their subjects, or to change the style of thought or the very form of a specific science.

In the remaining chapters of this book I apply, and participate in, politics of truth. I focus on two cases: Left-handers and Icelanders as subjects of science. In the fourth chapter I showed how the history of left-handedness has affected the scientific discourse on left-handedness and the troubles of the classification of left-handedness in scientific studies. I also sketched the relations of power that gave rise to scientific studies on left-handers and the relations of power that appeared while scientists tried to find an answer to the question of whether left-handers die on average younger than right-handers. Furthermore, I argued that left-handedness is not something obvious, it is not a clear and necessary category given by nature. Through that chapter, I showed how the production of knowledge about left-handers is incited by certain economic and political powers; how that knowledge is diffused and consumed through scientific literature, popular science literature and mass media; and how that knowledge has given rise to confrontation and conflict. I focused on the study of the longevity of left-handers as an example of the scientific discourse on left-handedness. I identified the power effects of that discourse and argued that they could and should be resisted.

In the fifth chapter I discussed the scientific study of Icelanders, in the context of plans to establish a database containing health data, genetic data and genealogical data for almost all the population, turning it into a "totally informative population" for genetics research and a totally transparent population for public health. After describing the plans for the Icelandic database project, I argued that they have parallel ideas in the discourse of eugenics in Iceland, and that the database plans share many discursive features with the eugenics ideas. I tried to show that the scientific authority, which is applied to manage the Icelandic population as a population of docile research subjects, could be reduced by placing the scientific discourse in the context of the disqualified discourse of eugenics. Finally, I described how the scientific discourse of modern genetics turns Icelanders into a population of laboratory animals, a "totally informative population" of research subjects, which can be constantly observed by science—a sort of scientific panopticon. I argued that by becoming aware of these power effects of science one is already beginning to resist them. As in the case of left-handers, I applied politics of truth as an alternative to conventional epistemological and ethical critiques of science.

Printed in the United States
By Bookmasters